胶州湾生态系统长期变化图集
（2010—2015）

孙晓霞　孙松　主编

ATLAS OF LONG-TERM CHANGES IN THE JIAOZHOU BAY ECOSYSTEM (2010–2015)

海洋出版社

2023年·北京

图书在版编目(CIP)数据

胶州湾生态系统长期变化图集: 2010—2015 / 孙晓霞, 孙松主编. —北京: 海洋出版社, 2023.4

ISBN 978-7-5210-1037-4

Ⅰ.①胶… Ⅱ.①孙… ②孙… Ⅲ.①黄海–海湾–生态系–图集 Ⅳ.① X321.25-64

中国版本图书馆 CIP 数据核字 (2022) 第 209442 号

胶州湾生态系统长期变化图集 (2010—2015)
JIAOZHOUWAN SHENGTAI XITONG CHANGQI BIANHUA TUJI (2010-2015)

审图号：鲁 5G (2022) 031 号

责任编辑：苏　勤
责任印制：安　淼

海洋出版社 出版发行
http://www.oceanpress.com.cn
北京市海淀区大慧寺路 8 号　邮编：100081
鸿博昊天科技有限公司印刷　新华书店北京发行所经销
2023 年 4 月第 1 版　2023 年 4 月第 1 次印刷
开本：889 mm × 1194 mm　1/16　印张：26.5
字数：180 千字　定价：298.00 元
发行部：010-62100090　邮购部：010-62100072　总编室：010-62100034

海洋版图书印、装错误可随时退换

EDITORIAL BOARD
编委会

主　编：孙晓霞　孙　松
副主编：赵永芳　张光涛
绘　图：赵永芳　赵辰浩　张康宁　杨适萌

Editors-in-chief: Sun Xiaoxia Sun Song
Associate editors-in-chief: Zhao Yongfang Zhang Guangtao
Cartography: Zhao Yongfang Zhao Chenhao
　　　　　　　Zhang Kangning Yang Shimeng

前 言

海洋在人类经济、社会发展中占有越来越重要的地位。向海洋进军，开发海洋资源，拓展生存空间，发展海洋经济，是我国的重要战略决策，也是蓝色经济区建设的核心内容。

随着海洋经济的发展，大量人口向沿海地带聚集，沿海的工业化、城镇化建设、海洋岸线的开发、大量工业污水和生活污水的排放、陆源物质的排放、港口建设、跨海大桥的建设、围填海、大量的养殖活动等对海洋生态系统，特别是近海生态系统造成了严重的影响。如何实现协调和可持续发展是摆在我们面前的十分艰巨的任务。

维护生态系统健康、海洋生物资源可持续利用、海洋环境可持续发展是实现海洋经济战略的必经之路。要做到这一切的首要任务是深入研究海洋生态系统演变的规律，了解海洋中已经发生了哪些变化、现在正在发生什么变化，更重要的是预测未来会发生什么样的变化。对海洋生态系统长期变化研究的关键是相关信息的获取，因此海洋长期观测是十分必要的。由于海洋观测投入很大，因此如何实现已有数据和资料的共享显得非常重要。让已经获取的数据和资料充分发挥作用，让所有海洋科技工作者、学生和社会各界都能够方便地使用这些资料是大家共同的愿望。

海湾生态系统是与人类活动关系最密切的生态系统。以胶州湾为例，这里是青岛市的母亲湾，是一个集港口、工业、农业、海水养殖业、滨海旅游业于一体的我国东部发达地区的典型海湾。胶州湾生态系统不仅受到自然气候变化的影响，更受到人类活动的影响。胶州湾是开展人类活动和自然变化双重作用下海洋生态系统演变机理研究的理想场所，也是我国海湾生态系统中开展研究最多的区域。

山东胶州湾海洋生态系统国家野外科学观测研究站（以下简称胶州湾站）始建于1981年，原名为黄岛海水养殖试验场，1986年改名为黄岛增养殖实验站。中国生态系统研究网络（CERN）组建后，该站1991年成为CERN 29个野外观测基本站之一，改名为胶州湾海洋生态系统定位研究站，是CERN中代表我国温带海域唯一的集监测、研究与示范为一体的综合性生态系统长期研究站。胶州湾站于2005年被科技部批准成为国家生态系统野外科学观测研究站，正式命名为山东胶州湾海洋生态系统国家野外科学观测研究站。2019年在科技部国家野外站首次评估中获评优秀国家野外站。

建站之初，胶州湾站主要从事鱼、虾、贝工厂化育苗和高产养殖关键技术的研究、示范工作。出色地完成了"胶州湾海洋环境及资源调查和鱼虾种苗放流增殖实验"等一系列重大项目，我国海洋水产养殖中三次浪潮（海带养殖、对虾养殖和扇贝养殖）的兴起主要始于胶州湾。20世纪90年代以后，胶州湾站针对日渐突出的环境问题，开始对生态系统的结构与功能进行综合调查和长期监测。进入21世纪以来，胶州湾站开始从全球变化和人类活动影响的高度全面考量生态系统的动态变化，研究人与自然和谐发展的途径与关键技术。胶州湾站拥有超过30年的长期时间序列、综合观测数据，收集了100年的气象资料、50年的长期考察资料和生物样品。自2003年开始，将监测频率增加到每月1次，有长期观测站位14个，基本涵盖了胶州湾主要生态区域。

本图集是在胶州湾生态系统长期变化数据集的基础上完成的，也是在2011年出版的《胶州湾生态系统长期变化图集（1997—2009）》基础上的延续，以直观的方式，反映胶州湾生态系统主要要素的分布特征、季节变化、年际变化和长期变化。本图集的所有数据均是胶州湾站相关团队科研人员在胶州湾进行现场考察获取的。希望本图集能够为近海生态系统长期变化研究、海洋生态系统健康评估、海洋生态环境保护、海洋经济可持续发展提供基础资料和决策依据。

本图集是在一系列国家科研项目和机构的资助下完成的，包括中国科学院国际伙伴计划（133137KYSB20200002、121311KYSB20190029-3）、国家自然科学基金项目（U2006206）、中国科学院战略先导科技专项子课题（XDA19060204）、国家重点基础研究发展计划项目课题（2014CB441504）、科技部国家科技基础条件平台中心、中国生态系统研究网络（CERN）等。在此感谢中国科学院、科技部、国家自然科学基金委员会、青岛市和中国科学院海洋研究所给予的支持，感谢所有参与胶州湾生态系统观测的科学家和技术人员，特别感谢中国科学院海洋研究所"创新"号科学考察船的全体船员对胶州湾长期观测的支持，感谢山东胶州湾海洋生态系统国家野外科学观测研究站的各位科技人员的不懈努力，感谢中国科学院海洋研究所各位参与胶州湾生态系统研究和观测的科技工作者。

编　者

2022年2月于青岛

Foreword

The ocean plays an increasingly important role in economic and social development. Developing marine resources, expanding living space, and developing the marine economy is not only an important strategy in China, but is also the core of the construction of the blue economic zone.

Development of the marine economy results in large numbers of people gathering in coastal areas. The marine ecosystem, especially the coastal ecosystem, has been heavily influenced by an abundance of human activities, including industrialization, urbanization, development of the coastline, discharge of industrial effluent, domestic sewage and land-source pollution, harbor construction, bridge construction, reclamation, and aquaculture. Accordingly, determining methods to achieve coordinated, sustainable development is difficult.

Maintenance of ecosystem health, sustainable use of marine living resources, and sustainable development of the marine environment is the only way to realize a marine economic strategy. To achieve this goal, it is important to understand the evolution of the marine ecosystem, e.g., what changes have taken place, what happens now, and more importantly, what will happen in the future. To understand the long-term changes in the marine ecosystem, it is necessary to obtain relevant information; therefore, long-term ocean observations are essential. Because ocean observation is very expensive, it is important to share existing data and information. It is our common wish to share these data and information so that all marine scientists, students and the relevant community have easy access to this information.

Bay ecosystems are closely related to human activities. For example, Jiaozhou Bay is known as the mother bay of Qingdao. This bay combines ports, industry, agriculture, aquaculture and coastal tourism, acting as a typical bay in an eastern developed area. The Jiaozhou Bay ecosystem is not only influenced by natural climate change, but also by human activities. The bay is an ideal place to investigate the evolution of a marine ecosystem impacted by both human activities and natural changes. Accordingly, the Jiaozhou Bay ecosystem is also one of the most-studied bay ecosystems.

Jiaozhou Bay Marine Ecosystem Research Station, which was founded in 1981, was formerly known as the Huangdao testing ground for marine aquaculture. The name was changed to Huangdao Maricultural Experiment Station in 1986. After the Chinese Ecosystem Research Network (CERN) was established, the station became one of the 29 basic stations for field observations and was renamed the Jiaozhou Bay Marine Ecosystem Research Station. Jiaozhou Bay Station is the only comprehensive ecosystem research station in temperate coastalwaters in CERN, and it integrates monitoring, research and demonstration. The facility was approved by the Ministry of Science and Technology as a national observation station in 2005, at which time it was officially named the National Research Station of Jiaozhou Bay Marine Ecosystem, Shandong. In 2019, Jiaozhou Bay

Station was evaluated as the "Excellent Station" in the first round of National station evaluation launched by Ministry of Science and Technology.

When the station was first established, the research and demonstrations mainly focused on industrialized breeding and developing methods of high yield aquaculture of fish, shrimp and shellfish. A series of major projects have been completed at the station, including the Jiaozhou Bay marine environment and resources investigation and testing of the reproduction and release of fish and shrimp seed. The three waves of marine aquaculture in China (seaweed, shrimp and scallop culture) began in Jiaozhou Bay. Owing to the increasing environmental problems in the bay, the members of the Jiaozhou Bay station started a comprehensive survey and long term monitoring of the ecosystem structure and function during the 1990s. In the 21st century, the members of Jiaozhou Bay station fully considered the dynamic changes in the ecosystem by combining the multiple influences of global change and human activities to study approaches and key technologies designed to enable the harmonious development of humans and nature. Jiaozhou Bay station has more than 30 years of long term data series, comprehensive observation data, over 100 years of meteorological data, and 50 years of long-term study data and biological samples. Since 2003, the observation frequency has increased to once a month on the 14 long-term observation stations covering the major ecological regions of Jiaozhou Bay.

The atlas was generated based on the dataset of the long-term changes in the Jiaozhou Bay ecosystem. It is also a continuation of the *Atlas of Long-term Changes in the Jiaozhou Bay Ecosystem (1997−2009)* published in 2011. It reflects the distribution characteristics, seasonal changes, annual changes and long term changes in the major elements of the ecosystem in an intuitive manner. All data were obtained in situ by staff from the Jiaozhou Bay Station and the Institute of Oceanology, Chinese Academy of Sciences. We hope that this atlas provides fundamental data and a decision-making basis for research into long term changes in the coastal ecosystem, marine ecosystem health assessment, marine environment protection, and sustainable development of the marine economy.

The atlas was funded by a series of research grants and institutions, including those from International Partnership Program of Chinese Academy of Sciences (133137KYSB20200002, 121311KYSB20190029-3), the National Natural Science Foundation of China (U2006206), Strategic Priority Research Program of the Chinese Academy of Sciences (XDA19060204), the National Basic Research Program of China (No.2014CB441504), the Infrastructure Project of Ecosystem Network Observation and the data-sharing system of the Ministry of Science and Technology, and Chinese Ecosystem Research Network etc. We appreciate the financial support provided by the Chinese Academy of Sciences, Ministry of Science and Technology, National Natural Science Foundation of China, Qingdao Municipal Science and Technology Commission, and Institute of Oceanology of the Chinese Academy of Sciences. Additionally, we thank all of the scientists and the technical staff involved in the Jiaozhou Bay ecosystem observations, especially the crew of the "Innovation" research vessels for their support during the longterm observation of Jiaozhou Bay. We also appreciate the tireless efforts of the staff of the Jiaozhou Bay station as well as our colleagues at the Institute of Oceanology, Chinese Academy of Sciences for their participation in Jiaozhou Bay ecosystem research and observations.

Editors

February 2022, Qingdao

技术说明

本图集是在山东胶州湾海洋生态系统国家野外科学观测研究站对胶州湾生态系统长期观测与研究的基础上完成的。本图集选择代表物理海洋学、化学海洋学、生物海洋学、海洋地质等不同领域的30多项监测指标，在2010—2015年间12个观测站（JZB11、JZB14站除外，在跨海大桥建设期间无法到达采样）季度月（即2月、5月、8月、11月中旬）的系统调查数据，对表层监测要素进行插值绘图。

胶州湾常规调查利用"创新"号科学考察船协助完成，目前常规调查站位14个，调查监测项目50余项。本图集涉及的指标除了沉积物总磷、沉积物有机质含量外，其他监测指标的样品采集、保存、分析和质量控制参照海洋调查规范和海洋监测规范执行。沉积物样品先用蒸馏水过80目的筛绢冲洗，去掉贝壳等杂质，并脱盐，然后置于恒温干燥箱中烘干（60 ℃，4天）。用研钵将干燥的沉积物样品研碎，过80目筛绢待测。称取0.2 g左右干样于30 mm×50 mm称量瓶中，加2 mL Mg(NO$_3$)$_2$溶液 (0.1 M)，于95℃烘干后进行灰化（500 ℃，3小时），对灰化后的残渣用10 mL HCl溶液 (0.2 M) 于80℃浸提0.5小时，对浸提液离心，并将上清液转移到100 mL容量瓶中，定容至刻度，取50 mL用磷钼蓝分光光度法进行磷测定，即得到样品中的总磷含量（Zhou et al., 2002），有机质含量通过灰化前后的重量差计算得到。所有数据进一步遵循中国生态系统研究网络（CERN）统一规范的数据处理方法和质量控制体系进行质控。

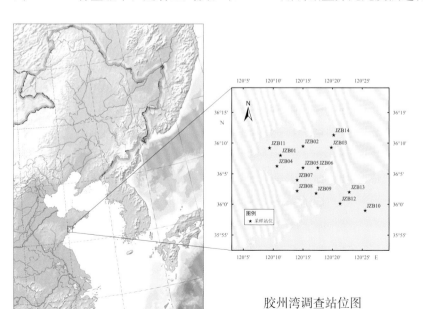

胶州湾调查站位图

胶州湾观测样品分析方法

分析项目	分析方法	参照国标/文献
温度	CTD	海洋调查规范 (GB/T 12763.2—2007)
盐度	CTD	海洋调查规范 (GB/T 12763.2—2007)
透明度	萨式盘法	海洋调查规范 (GB/T 12763.2—2007)
pH 值	pH 计法	海洋监测规范 (GB 17378.4—2007)
溶解氧	碘量法	海洋监测规范 (GB 17378.4—2007)
硅酸盐	硅钼蓝法	海洋监测规范 (GB 17378.4—2007)
磷酸盐	磷钼蓝法	海洋监测规范 (GB 17378.4—2007)
亚硝酸盐	重氮–偶氮法	海洋监测规范 (GB 17378.4—2007)
硝酸盐	镉铜还原后重氮–偶氮法	海洋监测规范 (GB 17378.4—2007)
铵盐	水杨酸钠法和靛酚兰法	海洋监测规范 (GB 17378.4—2007)
溶解有机碳	高温燃烧法	海洋监测规范 (GB 17378.4—2007)
化学需氧量	碱性高锰酸钾法	海洋监测规范 (GB 17378.4—2007)
总磷	过硫酸钾氧化法	海洋监测规范 (GB 17378.4—2007)
总氮	过硫酸钾氧化法	海洋监测规范 (GB 17378.4—2007)
颗粒有机碳	元素分析仪法	海洋调查规范 (GB/T 12763.4—2007)
悬浮体浓度	重量法	海洋监测规范 (GB 17378.4—2007)
沉积物含水率	重量法	海洋监测规范 (GB 17378.5—2007)
沉积物总磷	分光光度法	Zhou Y, Zhang F S, Yang H S. Extraction of phosphorus in natural waters and sediments by ignition method. Chinese Journal of Analytical Chemistry, 2002, 30(7): 861–864.
沉积物总氮	凯式滴定法	海洋监测规范 (GB 17378.5—2007)
沉积物砂土含量	Cilas 940L 型激光粒度仪	海洋调查规范 (GB/T 13909—1992)
沉积物粉砂含量	Cilas 940L 型激光粒度仪	海洋调查规范 (GB/T 13909—1992)
沉积物黏土含量	Cilas 940L 型激光粒度仪	海洋调查规范 (GB/T 13909—1992)
沉积物有机质含量	高温燃烧法	
叶绿素 a 浓度	荧光分光光度法	海洋监测规范 (GB 17378.7—2007)
浮游植物细胞数量	浓缩计数法	海洋监测规范 (GB 17378.7—2007)
浮游动物丰度	计数法	海洋监测规范 (GB 17378.7—2007)
浮游动物生物量	称重法	海洋监测规范 (GB 17378.7—2007)
水样含菌数	流式细胞仪法	海洋微微型光合浮游生物的测定 (GB/T 30737—2014)
大肠杆菌丰度	发酵法	海洋监测规范 (GB 17378.7—2007)
底栖生物总密度	计数法	海洋监测规范 (GB 17378.7—2007)
底栖生物总生物量	称重法	海洋监测规范 (GB 17378.7—2007)

Technical Specification

The atlas was produced based on the long term observation data of the Jiaozhou Bay ecosystem. More than 30 monitoring parameters included physical, chemical, biological and geological elements at 12 stations were targeted in the atlas. Data from stations JZB11 and JZB14 were not available, due to inaccessible sampling during the construction of the cross-sea bridge. The data for the surface layer, obtained quarterly in the middle of February, May, August and November between 2010 and 2015, were used for the interpolation mapping in the atlas.

"Innovation" research vessel was used for the long term observations in Jiaozhou Bay. More than 50 parameters were monitored at 14 stations. The sample collection, preservation, analysis and quality control of the parameters involved in the atlas were performed according to the national standards, the Specification for Marine Monitoring and the Specifications for Oceanographic Survey, except for total phosphorus and organic matter content in sediment. Sediment samples were rinsed with distilled water through 80 mesh sieve to remove shells and other impurities. The samples were desalted and then dried at 60 ℃ for 4 days. The dried sediment samples were ground with a mortar and filtered through 80 mesh sieve to be measured. Two millilitres Mg$(NO_3)_2$ (0.1 M) were added to 0.2 g of dry sediment in a 30 mm × 50 mm weighing flask and dried at 95℃. Sampled were ignited at 500℃ for 3 h, and the residue was extracted with 10 mL HCl (0.2 M) at 80℃ for 0.5 h. After centrifugation of the extracts, the supernatant was transferred to a 100 mL volumetric flask and fixed. The total phosphorus in sediment was obtained by taking 50 mL of the solution for phosphorus determination using phosphomolybdenum blue spectrophotometry (Zhou et. al., 2002). And the organic matter content in sediment was calculated by the weight difference before and after igniting. All data were further quality controlled according to the specifications of the China Ecosystem Research Network (CERN), including data processing methods and quality control system.

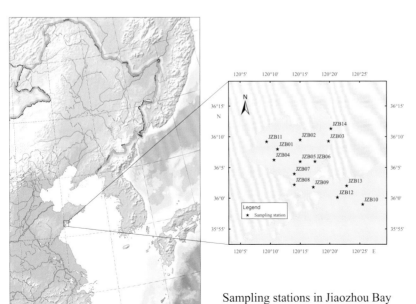

Sampling stations in Jiaozhou Bay

Analysis methods for parameters measured in the survey of Jiaozhou Bay

Parameters	Method	National standard of The Peoples Republic of China
Temperature	CTD	The specification for oceanographic survey (GB/T 12763.2-2007)
Salinity	CTD	The specification for oceanographic survey (GB/T 12763.2-2007)
Secchi disk depth	Secchi disk method	The specification for oceanographic survey (GB/T 12763.2-2007)
pH	pH meter	The specification for marine monitoring (GB 17378.4-2007)
Dissolved oxygen	Iodometric method	The specification for marine monitoring (GB 17378.4-2007)
Silicate	Silicon molybdenum blue method	The specification for marine monitoring (GB 17378.4-2007)
Phosphate	Phosphorus molybdenum blue method	The specification for marine monitoring (GB 17378.4-2007)
Nitrite	Diazo-coupling method	The specification for marine monitoring (GB 17378.4-2007)
Nitrate	Cadmium-copper column reduction	The specification for marine monitoring (GB 17378.4-2007)
Ammonium	Salicylic acid and Indophenol blue Spectrophotometry	The specification for marine monitoring (GB 17378.4-2007)
Dissolved organic carbon	High-temperature combustion method (HTC)	The specification for marine monitoring (GB 17378.4-2007)
Chemical oxygen demand	Basic potassium permanganate	The specification for marine monitoring (GB 17378.4-2007)
Total phosphorus	Alkaline potassium persulfate oxidation	The specification for marine monitoring (GB 17378.4-2007)
Total nitrogen	Alkaline potassium persulfate oxidation	The specification for marine monitoring (GB 17378.4-2007)
Particulate organic carbon	Elemental analyzer method	The specification for oceanographic survey (GB/T 12763.4-2007)
Suspended matter concentration	Gravimetric method	The specification for marine monitoring (GB 17378.4-2007)
Sediment water content	Gravimetric method	The specification for marine monitoring (GB 17378.5-2007)
Total phosphorus in sediment	Spectrophotometry	Zhou Y, Zhang F S, Yang H S. Extraction of phosphorus in natural waters and sediments by ignition method. Chinese Journal of Analytical Chemistry, 2002, 30(7):861−864.
Total nitrogen in sediment	Kjeldahl method	The specification for marine monitoring (GB 17378.5-2007)
Sand content in sediment	Cilas 940L laser particle size analyzer	The specification for oceanographic survey (GB/T 13909-1992)
Silt content in sediment	Cilas 940L laser particle size analyzer	The specification for oceanographic survey (GB/T 13909-1992)
Clay content in sediment	Cilas 940L laser particle size analyzer	The specification for oceanographic survey (GB/T 13909-1992)
Organic matter content in sediment	High-temperature combustion method (HTC)	
Chlorophyll a	Fluorescence spectrophotometry	The specification for marine monitoring (GB 17378.7-2007)
Cell density of phytoplankton	Microscopic counting method	The specification for marine monitoring (GB 17378.7-2007)
Abundance of zooplankton	Microscopic counting method	The specification for marine monitoring (GB 17378.7-2007)
Biomass of zooplankton	Gravimetric method	The specification for marine monitoring (GB 17378.7-2007)
Bacterial abundance	Flow cytometry	Determination of marine photosynthetic picoplankton (GB/T 30737-2014)
Coliform bacteria abundance	Fermentation	The specification for marine monitoring (GB 17378.7-2007)
Density of benthos	Count method	The specification for marine monitoring (GB 17378.7-2007)
Biomass of benthos	Gravimetric method	The specification for marine monitoring (GB 17378.7-2007)

目录

一、物理海洋 Physical Oceanography

1. 温度 Temperature ……………………………………………………………… 2

2010年2月温度分布 Distribution of temperature in February 2010 ……………… 2
2010年5月温度分布 Distribution of temperature in May 2010 …………………… 2
2010年8月温度分布 Distribution of temperature in August 2010 ………………… 3
2010年11月温度分布 Distribution of temperature in November 2010 …………… 3
2011年2月温度分布 Distribution of temperature in February 2011 ……………… 4
2011年5月温度分布 Distribution of temperature in May 2011 …………………… 4
2011年8月温度分布 Distribution of temperature in August 2011 ………………… 5
2011年11月温度分布 Distribution of temperature in November 2011 …………… 5
2012年2月温度分布 Distribution of temperature in February 2012 ……………… 6
2012年5月温度分布 Distribution of temperature in May 2012 …………………… 6
2012年8月温度分布 Distribution of temperature in August 2012 ………………… 7
2012年11月温度分布 Distribution of temperature in November 2012 …………… 7
2013年2月温度分布 Distribution of temperature in February 2013 ……………… 8
2013年5月温度分布 Distribution of temperature in May 2013 …………………… 8
2013年8月温度分布 Distribution of temperature in August 2013 ………………… 9
2013年11月温度分布 Distribution of temperature in November 2013 …………… 9
2014年2月温度分布 Distribution of temperature in February 2014 ……………… 10
2014年5月温度分布 Distribution of temperature in May 2014 …………………… 10
2014年8月温度分布 Distribution of temperature in August 2014 ………………… 11
2014年11月温度分布 Distribution of temperature in November 2014 …………… 11
2015年2月温度分布 Distribution of temperature in February 2015 ……………… 12
2015年5月温度分布 Distribution of temperature in May 2015 …………………… 12
2015年8月温度分布 Distribution of temperature in August 2015 ………………… 13
2015年11月温度分布 Distribution of temperature in November 2015 …………… 13

2. 盐度 Salinity …… 14

2010年2月盐度分布　Distribution of salinity in February 2010 …… 14
2010年5月盐度分布　Distribution of salinity in May 2010 …… 14
2010年8月盐度分布　Distribution of salinity in August 2010 …… 15
2010年11月盐度分布　Distribution of salinity in November 2010 …… 15
2011年2月盐度分布　Distribution of salinity in February 2011 …… 16
2011年5月盐度分布　Distribution of salinity in May 2011 …… 16
2011年8月盐度分布　Distribution of salinity in August 2011 …… 17
2011年11月盐度分布　Distribution of salinity in November 2011 …… 17
2012年2月盐度分布　Distribution of salinity in February 2012 …… 18
2012年5月盐度分布　Distribution of salinity in May 2012 …… 18
2012年8月盐度分布　Distribution of salinity in August 2012 …… 19
2012年11月盐度分布　Distribution of salinity in November 2012 …… 19
2013年2月盐度分布　Distribution of salinity in February 2013 …… 20
2013年5月盐度分布　Distribution of salinity in May 2013 …… 20
2013年8月盐度分布　Distribution of salinity in August 2013 …… 21
2013年11月盐度分布　Distribution of salinity in November 2013 …… 21
2014年2月盐度分布　Distribution of salinity in February 2014 …… 22
2014年5月盐度分布　Distribution of salinity in May 2014 …… 22
2014年8月盐度分布　Distribution of salinity in August 2014 …… 23
2014年11月盐度分布　Distribution of salinity in November 2014 …… 23
2015年2月盐度分布　Distribution of salinity in February 2015 …… 24
2015年5月盐度分布　Distribution of salinity in May 2015 …… 24
2015年8月盐度分布　Distribution of salinity in August 2015 …… 25
2015年11月盐度分布　Distribution of salinity in November 2015 …… 25

3. 透明度 Secchi disk depth …… 26

2010年2月透明度分布　Distribution of secchi disk depth in February 2010 …… 26
2010年5月透明度分布　Distribution of secchi disk depth in May 2010 …… 26
2010年8月透明度分布　Distribution of secchi disk depth in August 2010 …… 27
2010年11月透明度分布　Distribution of secchi disk depth in November 2010 …… 27
2011年2月透明度分布　Distribution of secchi disk depth in February 2011 …… 28
2011年5月透明度分布　Distribution of secchi disk depth in May 2011 …… 28
2011年8月透明度分布　Distribution of secchi disk depth in August 2011 …… 29
2011年11月透明度分布　Distribution of secchi disk depth in November 2011 …… 29
2012年2月透明度分布　Distribution of secchi disk depth in February 2012 …… 30
2012年5月透明度分布　Distribution of secchi disk depth in May 2012 …… 30
2012年8月透明度分布　Distribution of secchi disk depth in August 2012 …… 31
2012年11月透明度分布　Distribution of secchi disk depth in November 2012 …… 31
2013年2月透明度分布　Distribution of secchi disk depth in February 2013 …… 32

2013年5月透明度分布 Distribution of secchi disk depth in May 2013 ·········· 32
2013年8月透明度分布 Distribution of secchi disk depth in August 2013 ·········· 33
2013年11月透明度分布 Distribution of secchi disk depth in November 2013 ·········· 33
2014年2月透明度分布 Distribution of secchi disk depth in February 2014 ·········· 34
2014年5月透明度分布 Distribution of secchi disk depth in May 2014 ·········· 34
2014年8月透明度分布 Distribution of secchi disk depth in August 2014 ·········· 35
2014年11月透明度分布 Distribution of secchi disk depth in November 2014 ·········· 35
2015年2月透明度分布 Distribution of secchi disk depth in February 2015 ·········· 36
2015年5月透明度分布 Distribution of secchi disk depth in May 2015 ·········· 36
2015年8月透明度分布 Distribution of secchi disk depth in August 2015 ·········· 37
2015年11月透明度分布 Distribution of secchi disk depth in November 2015 ·········· 37

二、化学海洋　Chemical Oceanography

1. pH值 pH ·········· 40

2010年2月pH值分布 Distribution of pH in February 2010 ·········· 40
2010年5月pH值分布 Distribution of pH in May 2010 ·········· 40
2010年8月pH值分布 Distribution of pH in August 2010 ·········· 41
2010年11月pH值分布 Distribution of pH in November 2010 ·········· 41
2011年2月pH值分布 Distribution of pH in February 2011 ·········· 42
2011年5月pH值分布 Distribution of pH in May 2011 ·········· 42
2011年8月pH值分布 Distribution of pH in August 2011 ·········· 43
2011年11月pH值分布 Distribution of pH in November 2011 ·········· 43
2012年2月pH值分布 Distribution of pH in February 2012 ·········· 44
2012年5月pH值分布 Distribution of pH in May 2012 ·········· 44
2012年8月pH值分布 Distribution of pH in August 2012 ·········· 45
2012年11月pH值分布 Distribution of pH in November 2012 ·········· 45
2013年2月pH值分布 Distribution of pH in February 2013 ·········· 46
2013年5月pH值分布 Distribution of pH in May 2013 ·········· 46
2013年8月pH值分布 Distribution of pH in August 2013 ·········· 47
2013年11月pH值分布 Distribution of pH in November 2013 ·········· 47
2014年2月pH值分布 Distribution of pH in February 2014 ·········· 48
2014年5月pH值分布 Distribution of pH in May 2014 ·········· 48
2014年8月pH值分布 Distribution of pH in August 2014 ·········· 49
2014年11月pH值分布 Distribution of pH in November 2014 ·········· 49
2015年2月pH值分布 Distribution of pH in February 2015 ·········· 50
2015年5月pH值分布 Distribution of pH in May 2015 ·········· 50
2015年8月pH值分布 Distribution of pH in August 2015 ·········· 51
2015年11月pH值分布 Distribution of pH in November 2015 ·········· 51

2. 溶解氧 Dissolved oxygen …… 52

2010年2月溶解氧分布 Distribution of dissolved oxygen in February 2010 …… 52
2010年5月溶解氧分布 Distribution of dissolved oxygen in May 2010 …… 52
2010年8月溶解氧分布 Distribution of dissolved oxygen in August 2010 …… 53
2010年11月溶解氧分布 Distribution of dissolved oxygen in November 2010 …… 53
2011年2月溶解氧分布 Distribution of dissolved oxygen in February 2011 …… 54
2011年5月溶解氧分布 Distribution of dissolved oxygen in May 2011 …… 54
2011年8月溶解氧分布 Distribution of dissolved oxygen in August 2011 …… 55
2011年11月溶解氧分布 Distribution of dissolved oxygen in November 2011 …… 55
2012年2月溶解氧分布 Distribution of dissolved oxygen in February 2012 …… 56
2012年5月溶解氧分布 Distribution of dissolved oxygen in May 2012 …… 56
2012年8月溶解氧分布 Distribution of dissolved oxygen in August 2012 …… 57
2012年11月溶解氧分布 Distribution of dissolved oxygen in November 2012 …… 57
2013年2月溶解氧分布 Distribution of dissolved oxygen in February 2013 …… 58
2013年5月溶解氧分布 Distribution of dissolved oxygen in May 2013 …… 58
2013年8月溶解氧分布 Distribution of dissolved oxygen in August 2013 …… 59
2013年11月溶解氧分布 Distribution of dissolved oxygen in November 2013 …… 59
2014年2月溶解氧分布 Distribution of dissolved oxygen in February 2014 …… 60
2014年5月溶解氧分布 Distribution of dissolved oxygen in May 2014 …… 60
2014年8月溶解氧分布 Distribution of dissolved oxygen in August 2014 …… 61
2014年11月溶解氧分布 Distribution of dissolved oxygen in November 2014 …… 61
2015年2月溶解氧分布 Distribution of dissolved oxygen in February 2015 …… 62
2015年5月溶解氧分布 Distribution of dissolved oxygen in May 2015 …… 62
2015年8月溶解氧分布 Distribution of dissolved oxygen in August 2015 …… 63
2015年11月溶解氧分布 Distribution of dissolved oxygen in November 2015 …… 63

3. 硅酸盐 Silicate …… 64

2010年2月硅酸盐分布 Distribution of silicate in February 2010 …… 64
2010年5月硅酸盐分布 Distribution of silicate in May 2010 …… 64
2010年8月硅酸盐分布 Distribution of silicate in August 2010 …… 65
2010年11月硅酸盐分布 Distribution of silicate in November 2010 …… 65
2011年2月硅酸盐分布 Distribution of silicate in February 2011 …… 66
2011年5月硅酸盐分布 Distribution of silicate in May 2011 …… 66
2011年8月硅酸盐分布 Distribution of silicate in August 2011 …… 67
2011年11月硅酸盐分布 Distribution of silicate in November 2011 …… 67
2012年2月硅酸盐分布 Distribution of silicate in February 2012 …… 68
2012年5月硅酸盐分布 Distribution of silicate in May 2012 …… 68
2012年8月硅酸盐分布 Distribution of silicate in August 2012 …… 69
2012年11月硅酸盐分布 Distribution of silicate in November 2012 …… 69
2013年2月硅酸盐分布 Distribution of silicate in February 2013 …… 70

2013年5月硅酸盐分布　Distribution of silicate in May 2013	70
2013年8月硅酸盐分布　Distribution of silicate in August 2013	71
2013年11月硅酸盐分布　Distribution of silicate in November 2013	71
2014年2月硅酸盐分布　Distribution of silicate in February 2014	72
2014年5月硅酸盐分布　Distribution of silicate in May 2014	72
2014年8月硅酸盐分布　Distribution of silicate in August 2014	73
2014年11月硅酸盐分布　Distribution of silicate in November 2014	73
2015年2月硅酸盐分布　Distribution of silicate in February 2015	74
2015年5月硅酸盐分布　Distribution of silicate in May 2015	74
2015年8月硅酸盐分布　Distribution of silicate in August 2015	75
2015年11月硅酸盐分布　Distribution of silicate in November 2015	75

4. 磷酸盐 Phosphate　　76

2010年2月磷酸盐分布　Distribution of phosphate in February 2010	76
2010年5月磷酸盐分布　Distribution of phosphate in May 2010	76
2010年8月磷酸盐分布　Distribution of phosphate in August 2010	77
2010年11月磷酸盐分布　Distribution of phosphate in November 2010	77
2011年2月磷酸盐分布　Distribution of phosphate in February 2011	78
2011年5月磷酸盐分布　Distribution of phosphate in May 2011	78
2011年8月磷酸盐分布　Distribution of phosphate in August 2011	79
2011年11月磷酸盐分布　Distribution of phosphate in November 2011	79
2012年2月磷酸盐分布　Distribution of phosphate in February 2012	80
2012年5月磷酸盐分布　Distribution of phosphate in May 2012	80
2012年8月磷酸盐分布　Distribution of phosphate in August 2012	81
2012年11月磷酸盐分布　Distribution of phosphate in November 2012	81
2013年2月磷酸盐分布　Distribution of phosphate in February 2013	82
2013年5月磷酸盐分布　Distribution of phosphate in May 2013	82
2013年8月磷酸盐分布　Distribution of phosphate in August 2013	83
2013年11月磷酸盐分布　Distribution of phosphate in November 2013	83
2014年2月磷酸盐分布　Distribution of phosphate in February 2014	84
2014年5月磷酸盐分布　Distribution of phosphate in May 2014	84
2014年8月磷酸盐分布　Distribution of phosphate in August 2014	85
2014年11月磷酸盐分布　Distribution of phosphate in November 2014	85
2015年2月磷酸盐分布　Distribution of phosphate in February 2015	86
2015年5月磷酸盐分布　Distribution of phosphate in May 2015	86
2015年8月磷酸盐分布　Distribution of phosphate in August 2015	87
2015年11月磷酸盐分布　Distribution of phosphate in November 2015	87

5. 亚硝酸盐 Nitrite　　88

2010年2月亚硝酸盐分布　Distribution of nitrite in February 2010	88
2010年5月亚硝酸盐分布　Distribution of nitrite in May 2010	88

2010年8月亚硝酸盐分布 Distribution of nitrite in August 2010	89
2010年11月亚硝酸盐分布 Distribution of nitrite in November 2010	89
2011年2月亚硝酸盐分布 Distribution of nitrite in February 2011	90
2011年5月亚硝酸盐分布 Distribution of nitrite in May 2011	90
2011年8月亚硝酸盐分布 Distribution of nitrite in August 2011	91
2011年11月亚硝酸盐分布 Distribution of nitrite in November 2011	91
2012年2月亚硝酸盐分布 Distribution of nitrite in February 2012	92
2012年5月亚硝酸盐分布 Distribution of nitrite in May 2012	92
2012年8月亚硝酸盐分布 Distribution of nitrite in August 2012	93
2012年11月亚硝酸盐分布 Distribution of nitrite in November 2012	93
2013年2月亚硝酸盐分布 Distribution of nitrite in February 2013	94
2013年5月亚硝酸盐分布 Distribution of nitrite in May 2013	94
2013年8月亚硝酸盐分布 Distribution of nitrite in August 2013	95
2013年11月亚硝酸盐分布 Distribution of nitrite in November 2013	95
2014年2月亚硝酸盐分布 Distribution of nitrite in February 2014	96
2014年5月亚硝酸盐分布 Distribution of nitrite in May 2014	96
2014年8月亚硝酸盐分布 Distribution of nitrite in August 2014	97
2014年11月亚硝酸盐分布 Distribution of nitrite in November 2014	97
2015年2月亚硝酸盐分布 Distribution of nitrite in February 2015	98
2015年5月亚硝酸盐分布 Distribution of nitrite in May 2015	98
2015年8月亚硝酸盐分布 Distribution of nitrite in August 2015	99
2015年11月亚硝酸盐分布 Distribution of nitrite in November 2015	99

6. 硝酸盐 Nitrate … 100

2010年2月硝酸盐分布 Distribution of nitrate in February 2010	100
2010年5月硝酸盐分布 Distribution of nitrate in May 2010	100
2010年8月硝酸盐分布 Distribution of nitrate in August 2010	101
2010年11月硝酸盐分布 Distribution of nitrate in November 2010	101
2011年2月硝酸盐分布 Distribution of nitrate in February 2011	102
2011年5月硝酸盐分布 Distribution of nitrate in May 2011	102
2011年8月硝酸盐分布 Distribution of nitrate in August 2011	103
2011年11月硝酸盐分布 Distribution of nitrate in November 2011	103
2012年2月硝酸盐分布 Distribution of nitrate in February 2012	104
2012年5月硝酸盐分布 Distribution of nitrate in May 2012	104
2012年8月硝酸盐分布 Distribution of nitrate in August 2012	105
2012年11月硝酸盐分布 Distribution of nitrate in November 2012	105
2013年2月硝酸盐分布 Distribution of nitrate in February 2013	106
2013年5月硝酸盐分布 Distribution of nitrate in May 2013	106
2013年8月硝酸盐分布 Distribution of nitrate in August 2013	107
2013年11月硝酸盐分布 Distribution of nitrate in November 2013	107
2014年2月硝酸盐分布 Distribution of nitrate in February 2014	108
2014年5月硝酸盐分布 Distribution of nitrate in May 2014	108

2014年8月硝酸盐分布　Distribution of nitrate in August 2014 ……………………………………………… 109
2014年11月硝酸盐分布　Distribution of nitrate in November 2014 …………………………………………… 109
2015年2月硝酸盐分布　Distribution of nitrate in February 2015 …………………………………………… 110
2015年5月硝酸盐分布　Distribution of nitrate in May 2015 ………………………………………………… 110
2015年8月硝酸盐分布　Distribution of nitrate in August 2015 ……………………………………………… 111
2015年11月硝酸盐分布　Distribution of nitrate in November 2015 …………………………………………… 111

7. 铵盐 Ammonium ……………………………………………………………………… 112

2010年2月铵盐分布　Distribution of ammonium in February 2010 ………………………………………… 112
2010年5月铵盐分布　Distribution of ammonium in May 2010 ……………………………………………… 112
2010年8月铵盐分布　Distribution of ammonium in August 2010 …………………………………………… 113
2010年11月铵盐分布　Distribution of ammonium in November 2010 ……………………………………… 113
2011年2月铵盐分布　Distribution of ammonium in February 2011 ………………………………………… 114
2011年5月铵盐分布　Distribution of ammonium in May 2011 ……………………………………………… 114
2011年8月铵盐分布　Distribution of ammonium in August 2011 …………………………………………… 115
2011年11月铵盐分布　Distribution of ammonium in November 2011 ……………………………………… 115
2012年2月铵盐分布　Distribution of ammonium in February 2012 ………………………………………… 116
2012年5月铵盐分布　Distribution of ammonium in May 2012 ……………………………………………… 116
2012年8月铵盐分布　Distribution of ammonium in August 2012 …………………………………………… 117
2012年11月铵盐分布　Distribution of ammonium in November 2012 ……………………………………… 117
2013年2月铵盐分布　Distribution of ammonium in February 2013 ………………………………………… 118
2013年5月铵盐分布　Distribution of ammonium in May 2013 ……………………………………………… 118
2013年8月铵盐分布　Distribution of ammonium in August 2013 …………………………………………… 119
2013年11月铵盐分布　Distribution of ammonium in November 2013 ……………………………………… 119
2014年2月铵盐分布　Distribution of ammonium in February 2014 ………………………………………… 120
2014年5月铵盐分布　Distribution of ammonium in May 2014 ……………………………………………… 120
2014年8月铵盐分布　Distribution of ammonium in August 2014 …………………………………………… 121
2014年11月铵盐分布　Distribution of ammonium in November 2014 ……………………………………… 121
2015年2月铵盐分布　Distribution of ammonium in February 2015 ………………………………………… 122
2015年5月铵盐分布　Distribution of ammonium in May 2015 ……………………………………………… 122
2015年8月铵盐分布　Distribution of ammonium in August 2015 …………………………………………… 123
2015年11月铵盐分布　Distribution of ammonium in November 2015 ……………………………………… 123

8. 溶解有机碳 Dissolved organic carbon ……………………………………………… 124

2010年2月溶解有机碳分布　Distribution of dissolved organic carbon in February 2010 ………………… 124
2010年5月溶解有机碳分布　Distribution of dissolved organic carbon in May 2010 ……………………… 124
2010年8月溶解有机碳分布　Distribution of dissolved organic carbon in August 2010 …………………… 125
2010年11月溶解有机碳分布　Distribution of dissolved organic carbon in November 2010 ……………… 125
2011年2月溶解有机碳分布　Distribution of dissolved organic carbon in February 2011 ………………… 126
2011年5月溶解有机碳分布　Distribution of dissolved organic carbon in May 2011 ……………………… 126
2011年8月溶解有机碳分布　Distribution of dissolved organic carbon in August 2011 …………………… 127
2011年11月溶解有机碳分布　Distribution of dissolved organic carbon in November 2011 ……………… 127

2012年2月溶解有机碳分布 Distribution of dissolved organic carbon in February 2012 ……………… 128
2012年5月溶解有机碳分布 Distribution of dissolved organic carbon in May 2012 …………………… 128
2012年8月溶解有机碳分布 Distribution of dissolved organic carbon in August 2012 ………………… 129
2012年11月溶解有机碳分布 Distribution of dissolved organic carbon in November 2012 …………… 129
2013年2月溶解有机碳分布 Distribution of dissolved organic carbon in February 2013 ……………… 130
2013年5月溶解有机碳分布 Distribution of dissolved organic carbon in May 2013 …………………… 130
2013年8月溶解有机碳分布 Distribution of dissolved organic carbon in August 2013 ………………… 131
2013年11月溶解有机碳分布 Distribution of dissolved organic carbon in November 2013 …………… 131
2014年2月溶解有机碳分布 Distribution of dissolved organic carbon in February 2014 ……………… 132
2014年5月溶解有机碳分布 Distribution of dissolved organic carbon in May 2014 …………………… 132
2014年8月溶解有机碳分布 Distribution of dissolved organic carbon in August 2014 ………………… 133
2014年11月溶解有机碳分布 Distribution of dissolved organic carbon in November 2014 …………… 133
2015年2月溶解有机碳分布 Distribution of dissolved organic carbon in February 2015 ……………… 134
2015年5月溶解有机碳分布 Distribution of dissolved organic carbon in May 2015 …………………… 134
2015年8月溶解有机碳分布 Distribution of dissolved organic carbon in August 2015 ………………… 135
2015年11月溶解有机碳分布 Distribution of dissolved organic carbon in November 2015 …………… 135

9. 化学需氧量 Chemical oxygen demand ……………………………………………… 136

2010年2月化学需氧量分布 Distribution of chemical oxygen demand in February 2010 ……………… 136
2010年5月化学需氧量分布 Distribution of chemical oxygen demand in May 2010 …………………… 136
2010年8月化学需氧量分布 Distribution of chemical oxygen demand in August 2010 ………………… 137
2010年11月化学需氧量分布 Distribution of chemical oxygen demand in November 2010 …………… 137
2011年2月化学需氧量分布 Distribution of chemical oxygen demand in February 2011 ……………… 138
2011年5月化学需氧量分布 Distribution of chemical oxygen demand in May 2011 …………………… 138
2011年8月化学需氧量分布 Distribution of chemical oxygen demand in August 2011 ………………… 139
2011年11月化学需氧量分布 Distribution of chemical oxygen demand in November 2011 …………… 139
2012年2月化学需氧量分布 Distribution of chemical oxygen demand in February 2012 ……………… 140
2012年5月化学需氧量分布 Distribution of chemical oxygen demand in May 2012 …………………… 140
2012年8月化学需氧量分布 Distribution of chemical oxygen demand in August 2012 ………………… 141
2012年11月化学需氧量分布 Distribution of chemical oxygen demand in November 2012 …………… 141
2013年2月化学需氧量分布 Distribution of chemical oxygen demand in February 2013 ……………… 142
2013年5月化学需氧量分布 Distribution of chemical oxygen demand in May 2013 …………………… 142
2013年8月化学需氧量分布 Distribution of chemical oxygen demand in August 2013 ………………… 143
2013年11月化学需氧量分布 Distribution of chemical oxygen demand in November 2013 …………… 143
2014年2月化学需氧量分布 Distribution of chemical oxygen demand in February 2014 ……………… 144
2014年5月化学需氧量分布 Distribution of chemical oxygen demand in May 2014 …………………… 144
2014年8月化学需氧量分布 Distribution of chemical oxygen demand in August 2014 ………………… 145
2014年11月化学需氧量分布 Distribution of chemical oxygen demand in November 2014 …………… 145
2015年2月化学需氧量分布 Distribution of chemical oxygen demand in February 2015 ……………… 146
2015年5月化学需氧量分布 Distribution of chemical oxygen demand in May 2015 …………………… 146
2015年8月化学需氧量分布 Distribution of chemical oxygen demand in August 2015 ………………… 147
2015年11月化学需氧量分布 Distribution of chemical oxygen demand in November 2015 …………… 147

10. 总磷 Total phosphorus ········· 148

    2010年2月总磷分布　Distribution of total phosphorus in February 2010 ········· 148
    2010年5月总磷分布　Distribution of total phosphorus in May 2010 ········· 148
    2010年8月总磷分布　Distribution of total phosphorus in August 2010 ········· 149
    2010年11月总磷分布　Distribution of total phosphorus in November 2010 ········· 149
    2011年2月总磷分布　Distribution of total phosphorus in February 2011 ········· 150
    2011年5月总磷分布　Distribution of total phosphorus in May 2011 ········· 150
    2011年8月总磷分布　Distribution of total phosphorus in August 2011 ········· 151
    2011年11月总磷分布　Distribution of total phosphorus in November 2011 ········· 151
    2012年2月总磷分布　Distribution of total phosphorus in February 2012 ········· 152
    2012年5月总磷分布　Distribution of total phosphorus in May 2012 ········· 152
    2012年8月总磷分布　Distribution of total phosphorus in August 2012 ········· 153
    2012年11月总磷分布　Distribution of total phosphorus in November 2012 ········· 153
    2013年2月总磷分布　Distribution of total phosphorus in February 2013 ········· 154
    2013年5月总磷分布　Distribution of total phosphorus in May 2013 ········· 154
    2013年8月总磷分布　Distribution of total phosphorus in August 2013 ········· 155
    2013年11月总磷分布　Distribution of total phosphorus in November 2013 ········· 155
    2014年2月总磷分布　Distribution of total phosphorus in February 2014 ········· 156
    2014年5月总磷分布　Distribution of total phosphorus in May 2014 ········· 156
    2014年8月总磷分布　Distribution of total phosphorus in August 2014 ········· 157
    2014年11月总磷分布　Distribution of total phosphorus in November 2014 ········· 157
    2015年2月总磷分布　Distribution of total phosphorus in February 2015 ········· 158
    2015年5月总磷分布　Distribution of total phosphorus in May 2015 ········· 158
    2015年8月总磷分布　Distribution of total phosphorus in August 2015 ········· 159
    2015年11月总磷分布　Distribution of total phosphorus in November 2015 ········· 159

11. 总氮 Total nitrogen ········· 160

    2010年2月总氮分布　Distribution of total nitrogen in February 2010 ········· 160
    2010年5月总氮分布　Distribution of total nitrogen in May 2010 ········· 160
    2010年8月总氮分布　Distribution of total nitrogen in August 2010 ········· 161
    2010年11月总氮分布　Distribution of total nitrogen in November 2010 ········· 161
    2011年2月总氮分布　Distribution of total nitrogen in February 2011 ········· 162
    2011年5月总氮分布　Distribution of total nitrogen in May 2011 ········· 162
    2011年8月总氮分布　Distribution of total nitrogen in August 2011 ········· 163
    2011年11月总氮分布　Distribution of total nitrogen in November 2011 ········· 163
    2012年2月总氮分布　Distribution of total nitrogen in February 2012 ········· 164
    2012年5月总氮分布　Distribution of total nitrogen in May 2012 ········· 164
    2012年8月总氮分布　Distribution of total nitrogen in August 2012 ········· 165
    2012年11月总氮分布　Distribution of total nitrogen in November 2012 ········· 165

2013年2月总氮分布 Distribution of total nitrogen in February 2013 ·················· 166
2013年5月总氮分布 Distribution of total nitrogen in May 2013 ·················· 166
2013年8月总氮分布 Distribution of total nitrogen in August 2013 ·················· 167
2013年11月总氮分布 Distribution of total nitrogen in November 2013 ·················· 167
2014年2月总氮分布 Distribution of total nitrogen in February 2014 ·················· 168
2014年5月总氮分布 Distribution of total nitrogen in May 2014 ·················· 168
2014年8月总氮分布 Distribution of total nitrogen in August 2014 ·················· 169
2014年11月总氮分布 Distribution of total nitrogen in November 2014 ·················· 169
2015年2月总氮分布 Distribution of total nitrogen in February 2015 ·················· 170
2015年5月总氮分布 Distribution of total nitrogen in May 2015 ·················· 170
2015年8月总氮分布 Distribution of total nitrogen in August 2015 ·················· 171
2015年11月总氮分布 Distribution of total nitrogen in November 2015 ·················· 171

12. 颗粒有机碳 Particulate organic carbon ·················· 172

2010年2月颗粒有机碳分布 Distribution of particulate organic carbon in February 2010 ·················· 172
2010年5月颗粒有机碳分布 Distribution of particulate organic carbon in May 2010 ·················· 172
2010年8月颗粒有机碳分布 Distribution of particulate organic carbon in August 2010 ·················· 173
2011年2月颗粒有机碳分布 Distribution of particulate organic carbon in February 2011 ·················· 173
2011年5月颗粒有机碳分布 Distribution of particulate organic carbon in May 2011 ·················· 174
2011年8月颗粒有机碳分布 Distribution of particulate organic carbon in August 2011 ·················· 174
2011年11月颗粒有机碳分布 Distribution of particulate organic carbon in November 2011 ·················· 175
2012年2月颗粒有机碳分布 Distribution of particulate organic carbon in February 2012 ·················· 175
2012年5月颗粒有机碳分布 Distribution of particulate organic carbon in May 2012 ·················· 176
2012年8月颗粒有机碳分布 Distribution of particulate organic carbon in August 2012 ·················· 176
2012年11月颗粒有机碳分布 Distribution of particulate organic carbon in November 2012 ·················· 177
2013年2月颗粒有机碳分布 Distribution of particulate organic carbon in February 2013 ·················· 177
2013年5月颗粒有机碳分布 Distribution of particulate organic carbon in May 2013 ·················· 178
2013年8月颗粒有机碳分布 Distribution of particulate organic carbon in August 2013 ·················· 178
2013年11月颗粒有机碳分布 Distribution of particulate organic carbon in November 2013 ·················· 179
2014年2月颗粒有机碳分布 Distribution of particulate organic carbon in February 2014 ·················· 179
2014年5月颗粒有机碳分布 Distribution of particulate organic carbon in May 2014 ·················· 180
2014年8月颗粒有机碳分布 Distribution of particulate organic carbon in August 2014 ·················· 180
2014年11月颗粒有机碳分布 Distribution of particulate organic carbon in November 2014 ·················· 181
2015年2月颗粒有机碳分布 Distribution of particulate organic carbon in February 2015 ·················· 181
2015年5月颗粒有机碳分布 Distribution of particulate organic carbon in May 2015 ·················· 182
2015年8月颗粒有机碳分布 Distribution of particulate organic carbon in August 2015 ·················· 182
2015年11月颗粒有机碳分布 Distribution of particulate organic carbon in November 2015 ·················· 183

三、悬浮体和沉积物 Suspended Matter and Sediment

1. 悬浮体浓度 Suspended matter concentration ········ 186

  2010年2月悬浮体浓度分布 Distribution of suspended matter concentration in February 2010 ······ 186
  2010年5月悬浮体浓度分布 Distribution of suspended matter concentration in May 2010 ············ 186
  2010年8月悬浮体浓度分布 Distribution of suspended matter concentration in August 2010 ········· 187
  2010年11月悬浮体浓度分布 Distribution of suspended matter concentration in November 2010 ··· 187
  2011年2月悬浮体浓度分布 Distribution of suspended matter concentration in February 2011 ······ 188
  2011年5月悬浮体浓度分布 Distribution of suspended matter concentration in May 2011 ············ 188
  2011年8月悬浮体浓度分布 Distribution of suspended matter concentration in August 2011 ········· 189
  2011年11月悬浮体浓度分布 Distribution of suspended matter concentration in November 2011 ··· 189
  2012年2月悬浮体浓度分布 Distribution of suspended matter concentration in February 2012 ······ 190
  2012年5月悬浮体浓度分布 Distribution of suspended matter concentration in May 2012 ············ 190
  2012年8月悬浮体浓度分布 Distribution of suspended matter concentration in August 2012 ········· 191
  2012年11月悬浮体浓度分布 Distribution of suspended matter concentration in November 2012 ··· 191
  2013年2月悬浮体浓度分布 Distribution of suspended matter concentration in February 2013 ······ 192
  2013年5月悬浮体浓度分布 Distribution of suspended matter concentration in May 2013 ············ 192
  2013年8月悬浮体浓度分布 Distribution of suspended matter concentration in August 2013 ········· 193
  2013年11月悬浮体浓度分布 Distribution of suspended matter concentration in November 2013 ··· 193
  2014年2月悬浮体浓度分布 Distribution of suspended matter concentration in February 2014 ······ 194
  2014年5月悬浮体浓度分布 Distribution of suspended matter concentration in May 2014 ············ 194
  2014年8月悬浮体浓度分布 Distribution of suspended matter concentration in August 2014 ········· 195
  2014年11月悬浮体浓度分布 Distribution of suspended matter concentration in November 2014 ··· 195
  2015年2月悬浮体浓度分布 Distribution of suspended matter concentration in February 2015 ······ 196
  2015年5月悬浮体浓度分布 Distribution of suspended matter concentration in May 2015 ············ 196
  2015年8月悬浮体浓度分布 Distribution of suspended matter concentration in August 2015 ········· 197
  2015年11月悬浮体浓度分布 Distribution of suspended matter concentration in November 2015 ··· 197

2. 沉积物含水率 Sediment water content ········ 198

  2010年2月沉积物含水率分布 Distribution of sediment water content in February 2010 ·················· 198
  2010年5月沉积物含水率分布 Distribution of sediment water content in May 2010 ························ 198
  2010年8月沉积物含水率分布 Distribution of sediment water content in August 2010 ····················· 199
  2010年11月沉积物含水率分布 Distribution of sediment water content in November 2010 ················· 199
  2011年2月沉积物含水率分布 Distribution of sediment water content in February 2011 ·················· 200
  2011年5月沉积物含水率分布 Distribution of sediment water content in May 2011 ························ 200
  2011年8月沉积物含水率分布 Distribution of sediment water content in August 2011 ····················· 201
  2011年11月沉积物含水率分布 Distribution of sediment water content in November 2011 ················· 201
  2012年2月沉积物含水率分布 Distribution of sediment water content in February 2012 ·················· 202
  2012年5月沉积物含水率分布 Distribution of sediment water content in May 2012 ························ 202

2012年8月沉积物含水率分布 Distribution of sediment water content in August 2012 ·········· 203
2012年11月沉积物含水率分布 Distribution of sediment water content in November 2012 ·········· 203
2013年2月沉积物含水率分布 Distribution of sediment water content in February 2013 ·········· 204
2013年5月沉积物含水率分布 Distribution of sediment water content in May 2013 ·········· 204
2013年8月沉积物含水率分布 Distribution of sediment water content in August 2013 ·········· 205
2013年11月沉积物含水率分布 Distribution of sediment water content in November 2013 ·········· 205
2014年2月沉积物含水率分布 Distribution of sediment water content in February 2014 ·········· 206
2014年5月沉积物含水率分布 Distribution of sediment water content in May 2014 ·········· 206
2014年8月沉积物含水率分布 Distribution of sediment water content in August 2014 ·········· 207
2014年11月沉积物含水率分布 Distribution of sediment water content in November 2014 ·········· 207
2015年2月沉积物含水率分布 Distribution of sediment water content in February 2015 ·········· 208
2015年5月沉积物含水率分布 Distribution of sediment water content in May 2015 ·········· 208
2015年8月沉积物含水率分布 Distribution of sediment water content in August 2015 ·········· 209
2015年11月沉积物含水率分布 Distribution of sediment water content in November 2015 ·········· 209

3. 沉积物总磷 Total phosphorus in sediment ·········· 210

2010年2月沉积物总磷分布 Distribution of total phosphorus in sediment in February 2010 ·········· 210
2010年5月沉积物总磷分布 Distribution of total phosphorus in sediment in May 2010 ·········· 210
2010年8月沉积物总磷分布 Distribution of total phosphorus in sediment in August 2010 ·········· 211
2010年11月沉积物总磷分布 Distribution of total phosphorus in sediment in November 2010 ·········· 211
2011年2月沉积物总磷分布 Distribution of total phosphorus in sediment in February 2011 ·········· 212
2011年5月沉积物总磷分布 Distribution of total phosphorus in sediment in May 2011 ·········· 212
2011年8月沉积物总磷分布 Distribution of total phosphorus in sediment in August 2011 ·········· 213
2011年11月沉积物总磷分布 Distribution of total phosphorus in sediment in November 2011 ·········· 213
2012年2月沉积物总磷分布 Distribution of total phosphorus in sediment in February 2012 ·········· 214
2012年5月沉积物总磷分布 Distribution of total phosphorus in sediment in May 2012 ·········· 214
2012年8月沉积物总磷分布 Distribution of total phosphorus in sediment in August 2012 ·········· 215
2012年11月沉积物总磷分布 Distribution of total phosphorus in sediment in November 2012 ·········· 215
2013年2月沉积物总磷分布 Distribution of total phosphorus in sediment in February 2013 ·········· 216
2013年5月沉积物总磷分布 Distribution of total phosphorus in sediment in May 2013 ·········· 216
2013年8月沉积物总磷分布 Distribution of total phosphorus in sediment in August 2013 ·········· 217
2013年11月沉积物总磷分布 Distribution of total phosphorus in sediment in November 2013 ·········· 217
2014年2月沉积物总磷分布 Distribution of total phosphorus in sediment in February 2014 ·········· 218
2014年5月沉积物总磷分布 Distribution of total phosphorus in sediment in May 2014 ·········· 218
2014年8月沉积物总磷分布 Distribution of total phosphorus in sediment in August 2014 ·········· 219
2014年11月沉积物总磷分布 Distribution of total phosphorus in sediment in November 2014 ·········· 219
2015年2月沉积物总磷分布 Distribution of total phosphorus in sediment in February 2015 ·········· 220
2015年5月沉积物总磷分布 Distribution of total phosphorus in sediment in May 2015 ·········· 220
2015年8月沉积物总磷分布 Distribution of total phosphorus in sediment in August 2015 ·········· 221
2015年11月沉积物总磷分布 Distribution of total phosphorus in sediment in November 2015 ·········· 221

4. 沉积物总氮 Total nitrogen in sediment ... 222

2010年2月沉积物总氮分布　Distribution of total nitrogen in sediment in February 2010 ... 222
2010年5月沉积物总氮分布　Distribution of total nitrogen in sediment in May 2010 ... 222
2010年8月沉积物总氮分布　Distribution of total nitrogen in sediment in August 2010 ... 223
2010年11月沉积物总氮分布　Distribution of total nitrogen in sediment in November 2010 ... 223
2011年2月沉积物总氮分布　Distribution of total nitrogen in sediment in February 2011 ... 224
2011年5月沉积物总氮分布　Distribution of total nitrogen in sediment in May 2011 ... 224
2011年8月沉积物总氮分布　Distribution of total nitrogen in sediment in August 2011 ... 225
2011年11月沉积物总氮分布　Distribution of total nitrogen in sediment in November 2011 ... 225
2012年2月沉积物总氮分布　Distribution of total nitrogen in sediment in February 2012 ... 226
2012年5月沉积物总氮分布　Distribution of total nitrogen in sediment in May 2012 ... 226
2012年8月沉积物总氮分布　Distribution of total nitrogen in sediment in August 2012 ... 227
2012年11月沉积物总氮分布　Distribution of total nitrogen in sediment in November 2012 ... 227
2013年2月沉积物总氮分布　Distribution of total nitrogen in sediment in February 2013 ... 228
2013年5月沉积物总氮分布　Distribution of total nitrogen in sediment in May 2013 ... 228
2013年8月沉积物总氮分布　Distribution of total nitrogen in sediment in August 2013 ... 229
2013年11月沉积物总氮分布　Distribution of total nitrogen in sediment in November 2013 ... 229
2014年2月沉积物总氮分布　Distribution of total nitrogen in sediment in February 2014 ... 230
2014年5月沉积物总氮分布　Distribution of total nitrogen in sediment in May 2014 ... 230
2014年8月沉积物总氮分布　Distribution of total nitrogen in sediment in August 2014 ... 231
2014年11月沉积物总氮分布　Distribution of total nitrogen in sediment in November 2014 ... 231
2015年2月沉积物总氮分布　Distribution of total nitrogen in sediment in February 2015 ... 232
2015年5月沉积物总氮分布　Distribution of total nitrogen in sediment in May 2015 ... 232
2015年8月沉积物总氮分布　Distribution of total nitrogen in sediment in August 2015 ... 233
2015年11月沉积物总氮分布　Distribution of total nitrogen in sediment in November 2015 ... 233

5. 沉积物砂土含量 Sand content in sediment ... 234

2010年2月沉积物砂土含量分布　Distribution of sand content in sediment in February 2010 ... 234
2010年8月沉积物砂土含量分布　Distribution of sand content in sediment in August 2010 ... 234
2011年2月沉积物砂土含量分布　Distribution of sand content in sediment in February 2011 ... 235
2011年8月沉积物砂土含量分布　Distribution of sand content in sediment in August 2011 ... 235
2012年2月沉积物砂土含量分布　Distribution of sand content in sediment in February 2012 ... 236
2012年8月沉积物砂土含量分布　Distribution of sand content in sediment in August 2012 ... 236
2013年2月沉积物砂土含量分布　Distribution of sand content in sediment in February 2013 ... 237
2013年8月沉积物砂土含量分布　Distribution of sand content in sediment in August 2013 ... 237
2014年2月沉积物砂土含量分布　Distribution of sand content in sediment in February 2014 ... 238
2014年8月沉积物砂土含量分布　Distribution of sand content in sediment in August 2014 ... 238
2015年2月沉积物砂土含量分布　Distribution of sand content in sediment in February 2015 ... 239
2015年8月沉积物砂土含量分布　Distribution of sand content in sediment in August 2015 ... 239

6. 沉积物粉砂含量 Silt content in sediment 240

    2010年2月沉积物粉砂含量分布　Distribution of silt content in sediment in February 2010 240
    2010年8月沉积物粉砂含量分布　Distribution of silt content in sediment in August 2010 240
    2011年2月沉积物粉砂含量分布　Distribution of silt content in sediment in February 2011 241
    2011年8月沉积物粉砂含量分布　Distribution of silt content in sediment in August 2011 241
    2012年2月沉积物粉砂含量分布　Distribution of silt content in sediment in February 2012 242
    2012年8月沉积物粉砂含量分布　Distribution of silt content in sediment in August 2012 242
    2013年2月沉积物粉砂含量分布　Distribution of silt content in sediment in February 2013 243
    2013年8月沉积物粉砂含量分布　Distribution of silt content in sediment in August 2013 243
    2014年2月沉积物粉砂含量分布　Distribution of silt content in sediment in February 2014 244
    2014年8月沉积物粉砂含量分布　Distribution of silt content in sediment in August 2014 244
    2015年2月沉积物粉砂含量分布　Distribution of silt content in sediment in February 2015 245
    2015年8月沉积物粉砂含量分布　Distribution of silt content in sediment in August 2015 245

7. 沉积物黏土含量 Clay content in sediment 246

    2010年2月沉积物黏土含量分布　Distribution of clay content in sediment in February 2010 246
    2010年8月沉积物黏土含量分布　Distribution of clay content in sediment in August 2010 246
    2011年2月沉积物黏土含量分布　Distribution of clay content in sediment in February 2011 247
    2011年8月沉积物黏土含量分布　Distribution of clay content in sediment in August 2011 247
    2012年2月沉积物黏土含量分布　Distribution of clay content in sediment in February 2012 248
    2012年8月沉积物黏土含量分布　Distribution of clay content in sediment in August 2012 248
    2013年2月沉积物黏土含量分布　Distribution of clay content in sediment in February 2013 249
    2013年8月沉积物黏土含量分布　Distribution of clay content in sediment in August 2013 249
    2014年2月沉积物黏土含量分布　Distribution of clay content in sediment in February 2014 250
    2014年8月沉积物黏土含量分布　Distribution of clay content in sediment in August 2014 250
    2015年2月沉积物黏土含量分布　Distribution of clay content in sediment in February 2015 251
    2015年8月沉积物黏土含量分布　Distribution of clay content in sediment in August 2015 251

8. 沉积物有机质含量 Organic matter content in sediment 252

    2010年2月沉积物有机质含量分布　Distribution of organic matter content in sediment in February 2010 ... 252
    2010年5月沉积物有机质含量分布　Distribution of organic matter content in sediment in May 2010 252
    2010年8月沉积物有机质含量分布　Distribution of organic matter content in sediment in August 2010 253
    2010年11月沉积物有机质含量分布　Distribution of organic matter content in sediment in November 2010 ... 253
    2011年2月沉积物有机质含量分布　Distribution of organic matter content in sediment in February 2011 ... 254
    2011年5月沉积物有机质含量分布　Distribution of organic matter content in sediment in May 2011 254
    2011年8月沉积物有机质含量分布　Distribution of organic matter content in sediment in August 2011 255
    2011年11月沉积物有机质含量分布　Distribution of organic matter content in sediment in November 2011 255
    2012年2月沉积物有机质含量分布　Distribution of organic matter content in sediment in February 2012 ... 256
    2012年5月沉积物有机质含量分布　Distribution of organic matter content in sediment in May 2012 256
    2012年8月沉积物有机质含量分布　Distribution of organic matter content in sediment in August 2012 257
    2012年11月沉积物有机质含量分布　Distribution of organic matter content in sediment in November 2012 ... 257

2013年2月沉积物有机质含量分布　Distribution of organic matter content in sediment in February 2013 ⋯258
2013年5月沉积物有机质含量分布　Distribution of organic matter content in sediment in May 2013 ⋯⋯⋯258
2013年8月沉积物有机质含量分布　Distribution of organic matter content in sediment in August 2013 ⋯⋯259
2013年11月沉积物有机质含量分布　Distribution of organic matter content in sediment in November 2013　259
2014年2月沉积物有机质含量分布　Distribution of organic matter content in sediment in February 2014 ⋯260
2014年5月沉积物有机质含量分布　Distribution of organic matter content in sediment in May 2014 ⋯⋯⋯260
2014年8月沉积物有机质含量分布　Distribution of organic matter content in sediment in August 2014 ⋯⋯261
2014年11月沉积物有机质含量分布　Distribution of organic matter content in sediment in November 2014　261
2015年2月沉积物有机质含量分布　Distribution of organic matter content in sediment in February 2015 ⋯262
2015年5月沉积物有机质含量分布　Distribution of organic matter content in sediment in May 2015 ⋯⋯⋯262
2015年8月沉积物有机质含量分布　Distribution of organic matter content in sediment in August 2015 ⋯⋯263
2015年11月沉积物有机质含量分布　Distribution of organic matter content in sediment in November 2015　263

四、生物海洋　Biological Oceanography

1. 叶绿素 a 浓度 Chlorophyll a concentration ⋯⋯⋯⋯⋯⋯⋯⋯⋯⋯⋯⋯⋯⋯⋯⋯⋯⋯⋯266

2010年2月叶绿素a浓度分布　Distribution of chlorophyll a concentration in February 2010 ⋯⋯⋯⋯⋯266
2010年5月叶绿素a浓度分布　Distribution of chlorophyll a concentration in May 2010 ⋯⋯⋯⋯⋯⋯⋯266
2010年8月叶绿素a浓度分布　Distribution of chlorophyll a concentration in August 2010 ⋯⋯⋯⋯⋯⋯267
2010年11月叶绿素a浓度分布　Distribution of chlorophyll a concentration in November 2010 ⋯⋯⋯⋯267
2011年2月叶绿素a浓度分布　Distribution of chlorophyll a concentration in February 2011 ⋯⋯⋯⋯⋯268
2011年5月叶绿素a浓度分布　Distribution of chlorophyll a concentration in May 2011 ⋯⋯⋯⋯⋯⋯⋯268
2011年8月叶绿素a浓度分布　Distribution of chlorophyll a concentration in August 2011 ⋯⋯⋯⋯⋯⋯269
2011年11月叶绿素a浓度分布　Distribution of chlorophyll a concentration in November 2011 ⋯⋯⋯⋯269
2012年2月叶绿素a浓度分布　Distribution of chlorophyll a concentration in February 2012 ⋯⋯⋯⋯⋯270
2012年5月叶绿素a浓度分布　Distribution of chlorophyll a concentration in May 2012 ⋯⋯⋯⋯⋯⋯⋯270
2012年8月叶绿素a浓度分布　Distribution of chlorophyll a concentration in August 2012 ⋯⋯⋯⋯⋯⋯271
2012年11月叶绿素a浓度分布　Distribution of chlorophyll a concentration in November 2012 ⋯⋯⋯⋯271
2013年2月叶绿素a浓度分布　Distribution of chlorophyll a concentration in February 2013 ⋯⋯⋯⋯⋯272
2013年5月叶绿素a浓度分布　Distribution of chlorophyll a concentration in May 2013 ⋯⋯⋯⋯⋯⋯⋯272
2013年8月叶绿素a浓度分布　Distribution of chlorophyll a concentration in August 2013 ⋯⋯⋯⋯⋯⋯273
2013年11月叶绿素a浓度分布　Distribution of chlorophyll a concentration in November 2013 ⋯⋯⋯⋯273
2014年2月叶绿素a浓度分布　Distribution of chlorophyll a concentration in February 2014 ⋯⋯⋯⋯⋯274
2014年5月叶绿素a浓度分布　Distribution of chlorophyll a concentration in May 2014 ⋯⋯⋯⋯⋯⋯⋯274
2014年8月叶绿素a浓度分布　Distribution of chlorophyll a concentration in August 2014 ⋯⋯⋯⋯⋯⋯275
2014年11月叶绿素a浓度分布　Distribution of chlorophyll a concentration in November 2014 ⋯⋯⋯⋯275
2015年2月叶绿素a浓度分布　Distribution of chlorophyll a concentration in February 2015 ⋯⋯⋯⋯⋯276
2015年5月叶绿素a浓度分布　Distribution of chlorophyll a concentration in May 2015 ⋯⋯⋯⋯⋯⋯⋯276
2015年8月叶绿素a浓度分布　Distribution of chlorophyll a concentration in August 2015 ⋯⋯⋯⋯⋯⋯277
2015年11月叶绿素a浓度分布　Distribution of chlorophyll a concentration in November 2015 ⋯⋯⋯⋯277

2. 浮游植物细胞数量 Cell density of phytoplankton ·········· 278

 2010年2月浮游植物细胞数量分布 Distribution of phytoplankton cell density in February 2010 ··· 278
 2010年5月浮游植物细胞数量分布 Distribution of phytoplankton cell density in May 2010 ········ 278
 2010年8月浮游植物细胞数量分布 Distribution of phytoplankton cell density in August 2010 ····· 279
 2010年11月浮游植物细胞数量分布 Distribution of phytoplankton cell density in November 2010 279
 2011年2月浮游植物细胞数量分布 Distribution of phytoplankton cell density in February 2011 ··· 280
 2011年5月浮游植物细胞数量分布 Distribution of phytoplankton cell density in May 2011 ········ 280
 2011年8月浮游植物细胞数量分布 Distribution of phytoplankton cell density in August 2011 ····· 281
 2011年11月浮游植物细胞数量分布 Distribution of phytoplankton cell density in November 2011 281
 2012年2月浮游植物细胞数量分布 Distribution of phytoplankton cell density in February 2012 ··· 282
 2012年5月浮游植物细胞数量分布 Distribution of phytoplankton cell density in May 2012 ········ 282
 2012年8月浮游植物细胞数量分布 Distribution of phytoplankton cell density in August 2012 ····· 283
 2012年11月浮游植物细胞数量分布 Distribution of phytoplankton cell density in November 2012 283
 2013年2月浮游植物细胞数量分布 Distribution of phytoplankton cell density in February 2013 ··· 284
 2013年5月浮游植物细胞数量分布 Distribution of phytoplankton cell density in May 2013 ········ 284
 2013年8月浮游植物细胞数量分布 Distribution of phytoplankton cell density in August 2013 ····· 285
 2013年11月浮游植物细胞数量分布 Distribution of phytoplankton cell density in November 2013 285
 2014年2月浮游植物细胞数量分布 Distribution of phytoplankton cell density in February 2014 ··· 286
 2014年5月浮游植物细胞数量分布 Distribution of phytoplankton cell density in May 2014 ········ 286
 2014年8月浮游植物细胞数量分布 Distribution of phytoplankton cell density in August 2014 ····· 287
 2014年11月浮游植物细胞数量分布 Distribution of phytoplankton cell density in November 2014 287
 2015年2月浮游植物细胞数量分布 Distribution of phytoplankton cell density in February 2015 ··· 288
 2015年5月浮游植物细胞数量分布 Distribution of phytoplankton cell density in May 2015 ········ 288
 2015年8月浮游植物细胞数量分布 Distribution of phytoplankton cell density in August 2015 ····· 289
 2015年11月浮游植物细胞数量分布 Distribution of phytoplankton cell density in November 2015 289

3. 浮游动物丰度 Abundance of zooplankton ·········· 290

 2010年2月浮游动物丰度分布 Distribution of zooplankton abundance in February 2010 ·············290
 2010年5月浮游动物丰度分布 Distribution of zooplankton abundance in May 2010 ··················290
 2010年8月浮游动物丰度分布 Distribution of zooplankton abundance in August 2010 ···············291
 2010年11月浮游动物丰度分布 Distribution of zooplankton abundance in November 2010 ············291
 2011年2月浮游动物丰度分布 Distribution of zooplankton abundance in February 2011·············292
 2011年5月浮游动物丰度分布 Distribution of zooplankton abundance in May 2011 ··················292
 2011年8月浮游动物丰度分布 Distribution of zooplankton abundance in August 2011 ···············293
 2011年11月浮游动物丰度分布 Distribution of zooplankton abundance in November 2011 ············293
 2012年2月浮游动物丰度分布 Distribution of zooplankton abundance in February 2012 ············294
 2012年5月浮游动物丰度分布 Distribution of zooplankton abundance in May 2012 ··················294
 2012年8月浮游动物丰度分布 Distribution of zooplankton abundance in August 2012 ···············295
 2012年11月浮游动物丰度分布 Distribution of zooplankton abundance in November 2012 ············295
 2013年2月浮游动物丰度分布 Distribution of zooplankton abundance in February 2013 ············296
 2013年5月浮游动物丰度分布 Distribution of zooplankton abundance in May 2013 ··················296
 2013年8月浮游动物丰度分布 Distribution of zooplankton abundance in August 2013 ···············297

2013年11月浮游动物丰度分布 Distribution of zooplankton abundance in November 2013 ·············· 297
2014年2月浮游动物丰度分布 Distribution of zooplankton abundance in February 2014 ·············· 298
2014年5月浮游动物丰度分布 Distribution of zooplankton abundance in May 2014 ·············· 298
2014年8月浮游动物丰度分布 Distribution of zooplankton abundance in August 2014 ·············· 299
2014年11月浮游动物丰度分布 Distribution of zooplankton abundance in November 2014 ·············· 299
2015年2月浮游动物丰度分布 Distribution of zooplankton abundance in February 2015 ·············· 300
2015年5月浮游动物丰度分布 Distribution of zooplankton abundance in May 2015 ·············· 300
2015年8月浮游动物丰度分布 Distribution of zooplankton abundance in August 2015 ·············· 301
2015年11月浮游动物丰度分布 Distribution of zooplankton abundance in November 2015 ·············· 301

4. 浮游动物生物量 Biomass of zooplankton ·············· 302

2010年2月浮游动物生物量分布 Distribution of zooplankton biomass in February 2010 ·············· 302
2010年5月浮游动物生物量分布 Distribution of zooplankton biomass in May 2010 ·············· 302
2010年8月浮游动物生物量分布 Distribution of zooplankton biomass in August 2010 ·············· 303
2010年11月浮游动物生物量分布 Distribution of zooplankton biomass in November 2010 ·············· 303
2011年2月浮游动物生物量分布 Distribution of zooplankton biomass in February 2011 ·············· 304
2011年5月浮游动物生物量分布 Distribution of zooplankton biomass in May 2011 ·············· 304
2011年8月浮游动物生物量分布 Distribution of zooplankton biomass in August 2011 ·············· 305
2011年11月浮游动物生物量分布 Distribution of zooplankton biomass in November 2011 ·············· 305
2012年2月浮游动物生物量分布 Distribution of zooplankton biomass in February 2012 ·············· 306
2012年5月浮游动物生物量分布 Distribution of zooplankton biomass in May 2012 ·············· 306
2012年8月浮游动物生物量分布 Distribution of zooplankton biomass in August 2012 ·············· 307
2012年11月浮游动物生物量分布 Distribution of zooplankton biomass in November 2012 ·············· 307
2013年2月浮游动物生物量分布 Distribution of zooplankton biomass in February 2013 ·············· 308
2013年5月浮游动物生物量分布 Distribution of zooplankton biomass in May 2013 ·············· 308
2013年8月浮游动物生物量分布 Distribution of zooplankton biomass in August 2013 ·············· 309
2013年11月浮游动物生物量分布 Distribution of zooplankton biomass in November 2013 ·············· 309
2014年2月浮游动物生物量分布 Distribution of zooplankton biomass in February 2014 ·············· 310
2014年5月浮游动物生物量分布 Distribution of zooplankton biomass in May 2014 ·············· 310
2014年8月浮游动物生物量分布 Distribution of zooplankton biomass in August 2014 ·············· 311
2014年11月浮游动物生物量分布 Distribution of zooplankton biomass in November 2014 ·············· 311
2015年2月浮游动物生物量分布 Distribution of zooplankton biomass in February 2015 ·············· 312
2015年5月浮游动物生物量分布 Distribution of zooplankton biomass in May 2015 ·············· 312
2015年8月浮游动物生物量分布 Distribution of zooplankton biomass in August 2015 ·············· 313
2015年11月浮游动物生物量分布 Distribution of zooplankton biomass in November 2015 ·············· 313

5. 浮游动物优势种组成 Dominant species composition of zooplankton ·············· 314

2010年2月浮游动物优势种组成分布 Composition distribution of dominant zooplankton species in February 2010 ··· 314
2010年5月浮游动物优势种组成分布 Composition distribution of dominant zooplankton species in May 2010 ······· 314
2010年8月浮游动物优势种组成分布 Composition distribution of dominant zooplankton species in August 2010 ····· 315
2010年11月浮游动物优势种组成分布 Composition distribution of dominant zooplankton species in November 2010 ·· 315
2011年2月浮游动物优势种组成分布 Composition distribution of dominant zooplankton species in February 2011 ··· 316
2011年5月浮游动物优势种组成分布 Composition distribution of dominant zooplankton species in May 2011 ········ 316

2011年8月浮游动物优势种组成分布 Composition distribution of dominant zooplankton species in August 2011 …… 317
2011年11月浮游动物优势种组成分布 Composition distribution of dominant zooplankton species in November 2011 … 317
2012年2月浮游动物优势种组成分布 Composition distribution of dominant zooplankton species in February 2012 … 318
2012年5月浮游动物优势种组成分布 Composition distribution of dominant zooplankton species in May 2012 ……… 318
2012年8月浮游动物优势种组成分布 Composition distribution of dominant zooplankton species in August 2012 …… 319
2012年11月浮游动物优势种组成分布 Composition distribution of dominant zooplankton species in November 2012 … 319
2013年2月浮游动物优势种组成分布 Composition distribution of dominant zooplankton species in February 2013 … 320
2013年5月浮游动物优势种组成分布 Composition distribution of dominant zooplankton species in May 2013 ……… 320
2013年8月浮游动物优势种组成分布 Composition distribution of dominant zooplankton species in August 2013 …… 321
2013年11月浮游动物优势种组成分布 Composition distribution of dominant zooplankton species in November 2013 … 321
2014年2月浮游动物优势种组成分布 Composition distribution of dominant zooplankton species in February 2014 … 322
2014年5月浮游动物优势种组成分布 Composition distribution of dominant zooplankton species in May 2014 ……… 320
2014年8月浮游动物优势种组成分布 Composition distribution of dominant zooplankton species in August 2014 …… 323
2014年11月浮游动物优势种组成分布 Composition distribution of dominant zooplankton species in November 2014 … 323
2015年2月浮游动物优势种组成分布 Composition distribution of dominant zooplankton species in February 2015 … 324
2015年5月浮游动物优势种组成分布 Composition distribution of dominant zooplankton species in May 2015 ……… 324
2015年8月浮游动物优势种组成分布 Composition distribution of dominant zooplankton species in August 2015 …… 325
2015年11月浮游动物优势种组成分布 Composition distribution of dominant zooplankton species in November 2015 … 325

6. 水样含菌数 Bacterial abundance …… 326

2010年2月水样含菌数分布 Distribution of bacterial abundance in February 2010 …… 326
2010年5月水样含菌数分布 Distribution of bacterial abundance in May 2010 …… 326
2010年8月水样含菌数分布 Distribution of bacterial abundance in August 2010 …… 327
2010年11月水样含菌数分布 Distribution of bacterial abundance in November 2010 …… 327
2011年2月水样含菌数分布 Distribution of bacterial abundance in February 2011 …… 328
2011年5月水样含菌数分布 Distribution of bacterial abundance in May 2011 …… 328
2011年8月水样含菌数分布 Distribution of bacterial abundance in August 2011 …… 329
2011年11月水样含菌数分布 Distribution of bacterial abundance in November 2011 …… 329
2012年2月水样含菌数分布 Distribution of bacterial abundance in February 2012 …… 330
2012年5月水样含菌数分布 Distribution of bacterial abundance in May 2012 …… 330
2012年8月水样含菌数分布 Distribution of bacterial abundance in August 2012 …… 331
2012年11月水样含菌数分布 Distribution of bacterial abundance in November 2012 …… 331
2013年2月水样含菌数分布 Distribution of bacterial abundance in February 2013 …… 332
2013年5月水样含菌数分布 Distribution of bacterial abundance in May 2013 …… 332
2013年8月水样含菌数分布 Distribution of bacterial abundance in August 2013 …… 333
2013年11月水样含菌数分布 Distribution of bacterial abundance in November 2013 …… 333
2014年2月水样含菌数分布 Distribution of bacterial abundance in February 2014 …… 334
2014年5月水样含菌数分布 Distribution of bacterial abundance in May 2014 …… 334
2014年8月水样含菌数分布 Distribution of bacterial abundance in August 2014 …… 335
2014年11月水样含菌数分布 Distribution of bacterial abundance in November 2014 …… 335
2015年2月水样含菌数分布 Distribution of bacterial abundance in February 2015 …… 336
2015年5月水样含菌数分布 Distribution of bacterial abundance in May 2015 …… 336
2015年8月水样含菌数分布 Distribution of bacterial abundance in August 2015 …… 337
2015年11月水样含菌数分布 Distribution of bacterial abundance in November 2015 …… 337

7. 蓝细菌数 Cyanobacterial abundance ··············338

2010年2月蓝细菌数分布 Distribution of cyanobacterial abundance in February 2010··············338
2010年5月蓝细菌数分布 Distribution of cyanobacterial abundance in May 2010 ··············338
2010年8月蓝细菌数分布 Distribution of cyanobacterial abundance in August 2010 ··············339
2010年11月蓝细菌数分布 Distribution of cyanobacterial abundance in November 2010··············339
2011年2月蓝细菌数分布 Distribution of cyanobacterial abundance in February 2011 ··············340
2011年5月蓝细菌数分布 Distribution of cyanobacterial abundance in May 2011 ··············340
2011年8月蓝细菌数分布 Distribution of cyanobacterial abundance in August 2011 ··············341
2011年11月蓝细菌数分布 Distribution of cyanobacterial abundance in November 2011··············341
2012年2月蓝细菌数分布 Distribution of cyanobacterial abundance in February 2012··············342
2012年5月蓝细菌数分布 Distribution of cyanobacterial abundance in May 2012 ··············342
2012年8月蓝细菌数分布 Distribution of cyanobacterial abundance in August 2012 ··············343
2012年11月蓝细菌数分布 Distribution of cyanobacterial abundance in November 2012··············343
2013年2月蓝细菌数分布 Distribution of cyanobacterial abundance in February 2013··············344
2013年5月蓝细菌数分布 Distribution of cyanobacterial abundance in May 2013 ··············344
2013年8月蓝细菌数分布 Distribution of cyanobacterial abundance in August 2013 ··············345
2013年11月蓝细菌数分布 Distribution of cyanobacterial abundance in November 2013··············345
2014年2月蓝细菌数分布 Distribution of cyanobacterial abundance in February 2014··············346
2014年5月蓝细菌数分布 Distribution of cyanobacterial abundance in May 2014 ··············346
2014年8月蓝细菌数分布 Distribution of cyanobacterial abundance in August 2014 ··············347
2014年11月蓝细菌数分布 Distribution of cyanobacterial abundance in November 2014··············347
2015年2月蓝细菌数分布 Distribution of cyanobacterial abundance in February 2015··············348
2015年5月蓝细菌数分布 Distribution of cyanobacterial abundance in May 2015 ··············348
2015年8月蓝细菌数分布 Distribution of cyanobacterial abundance in August 2015 ··············349
2015年11月蓝细菌数分布 Distribution of cyanobacterial abundance in November 2015··············349

8. 大肠杆菌丰度 Coliform bacteria abundance ··············350

2010年2月大肠杆菌丰度分布 Distribution of coliform bacteria abundance in February 2010··········350
2010年5月大肠杆菌丰度分布 Distribution of coliform bacteria abundance in May 2010 ··············350
2010年8月大肠杆菌丰度分布 Distribution of coliform bacteria abundance in August 2010 ··············351
2010年11月大肠杆菌丰度分布 Distribution of coliform bacteria abundance in November 2010······351
2011年2月大肠杆菌丰度分布 Distribution of coliform bacteria abundance in February 2011 ··············352
2011年5月大肠杆菌丰度分布 Distribution of coliform bacteria abundance in May 2011 ··············352
2011年8月大肠杆菌丰度分布 Distribution of coliform bacteria abundance in August 2011 ··············353
2011年11月大肠杆菌丰度分布 Distribution of coliform bacteria abundance in November 2011······353
2012年2月大肠杆菌丰度分布 Distribution of coliform bacteria abundance in February 2012··············354
2012年5月大肠杆菌丰度分布 Distribution of coliform bacteria abundance in May 2012 ··············354
2012年8月大肠杆菌丰度分布 Distribution of coliform bacteria abundance in August 2012 ··············355
2012年11月大肠杆菌丰度分布 Distribution of coliform bacteria abundance in November 2012······355
2013年2月大肠杆菌丰度分布 Distribution of coliform bacteria abundance in February 2013··········356

2013年5月大肠杆菌丰度分布 Distribution of coliform bacteria abundance in May 2013 ············ 356
2013年8月大肠杆菌丰度分布 Distribution of coliform bacteria abundance in August 2013 ········· 357
2013年11月大肠杆菌丰度分布 Distribution of coliform bacteria abundance in November 2013······ 357
2014年2月大肠杆菌丰度分布 Distribution of coliform bacteria abundance in February 2014 ········ 358
2014年5月大肠杆菌丰度分布 Distribution of coliform bacteria abundance in May 2014 ············ 358
2014年8月大肠杆菌丰度分布 Distribution of coliform bacteria abundance in August 2014 ·········· 359
2014年11月大肠杆菌丰度分布 Distribution of coliform bacteria abundance in November 2014······ 359
2015年2月大肠杆菌丰度分布 Distribution of coliform bacteria abundance in February 2015 ········ 360
2015年5月大肠杆菌丰度分布 Distribution of coliform bacteria abundance in May 2015 ············ 360
2015年8月大肠杆菌丰度分布 Distribution of coliform bacteria abundance in August 2015 ·········· 361
2015年11月大肠杆菌丰度分布 Distribution of coliform bacteria abundance in November 2015······ 361

9. 底栖生物总生物量 Total biomass of benthos ·· 362

2010年2月底栖生物总生物量分布 Distribution of total benthic biomass in February 2010 ················ 362
2010年5月底栖生物总生物量分布 Distribution of total benthic biomass in May 2010 ···················· 362
2010年8月底栖生物总生物量分布 Distribution of total benthic biomass in August 2010 ·················· 363
2010年11月底栖生物总生物量分布 Distribution of total benthic biomass in November 2010 ·············· 363
2011年2月底栖生物总生物量分布 Distribution of total benthic biomass in February 2011················ 364
2011年5月底栖生物总生物量分布 Distribution of total benthic biomass in May 2011 ···················· 364
2011年8月底栖生物总生物量分布 Distribution of total benthic biomass in August 2011 ·················· 365
2011年11月底栖生物总生物量分布 Distribution of total benthic biomass in November 2011 ·············· 365
2012年2月底栖生物总生物量分布 Distribution of total benthic biomass in February 2012 ················ 366
2012年5月底栖生物总生物量分布 Distribution of total benthic biomass in May 2012 ···················· 366
2012年8月底栖生物总生物量分布 Distribution of total benthic biomass in August 2012 ·················· 367
2012年11月底栖生物总生物量分布 Distribution of total benthic biomass in November 2012 ·············· 367
2013年2月底栖生物总生物量分布 Distribution of total benthic biomass in February 2013 ················ 368
2013年5月底栖生物总生物量分布 Distribution of total benthic biomass in May 2013 ···················· 368
2013年8月底栖生物总生物量分布 Distribution of total benthic biomass in August 2013 ·················· 369
2013年11月底栖生物总生物量分布 Distribution of total benthic biomass in November 2013 ·············· 369
2014年2月底栖生物总生物量分布 Distribution of total benthic biomass in February 2014 ················ 370
2014年5月底栖生物总生物量分布 Distribution of total benthic biomass in May 2014 ···················· 370
2014年8月底栖生物总生物量分布 Distribution of total benthic biomass in August 2014 ·················· 371
2014年11月底栖生物总生物量分布 Distribution of total benthic biomass in November 2014 ·············· 371
2015年2月底栖生物总生物量分布 Distribution of total benthic biomass in February 2015 ················ 372
2015年5月底栖生物总生物量分布 Distribution of total benthic biomass in May 2015 ···················· 372
2015年8月底栖生物总生物量分布 Distribution of total benthic biomass in August 2015 ·················· 373
2015年11月底栖生物总生物量分布 Distribution of total benthic biomass in November 2015 ·············· 373

10. 底栖生物总密度 Total density of benthos ··· 374

   2010年2月底栖生物总密度分布　Distribution of total benthic density in February 2010 ··············· 374
   2010年5月底栖生物总密度分布　Distribution of total benthic density in May 2010 ······················ 374
   2010年8月底栖生物总密度分布　Distribution of total benthic density in August 2010 ··················· 375
   2010年11月底栖生物总密度分布　Distribution of total benthic density in November 2010 ··············· 375
   2011年2月底栖生物总密度分布　Distribution of total benthic density in February 2011 ··············· 376
   2011年5月底栖生物总密度分布　Distribution of total benthic density in May 2011 ······················ 376
   2011年8月底栖生物总密度分布　Distribution of total benthic density in August 2011 ··················· 377
   2011年11月底栖生物总密度分布　Distribution of total benthic density in November 2011 ··············· 377
   2012年2月底栖生物总密度分布　Distribution of total benthic density in February 2012 ··············· 378
   2012年5月底栖生物总密度分布　Distribution of total benthic density in May 2012 ······················ 378
   2012年8月底栖生物总密度分布　Distribution of total benthic density in August 2012 ··················· 379
   2012年11月底栖生物总密度分布　Distribution of total benthic density in November 2012 ··············· 379
   2013年2月底栖生物总密度分布　Distribution of total benthic density in February 2013 ··············· 380
   2013年5月底栖生物总密度分布　Distribution of total benthic density in May 2013 ······················ 380
   2013年8月底栖生物总密度分布　Distribution of total benthic density in August 2013 ··················· 381
   2013年11月底栖生物总密度分布　Distribution of total benthic density in November 2013 ··············· 381
   2014年2月底栖生物总密度分布　Distribution of total benthic density in February 2014 ··············· 382
   2014年5月底栖生物总密度分布　Distribution of total benthic density in May 2014 ······················ 382
   2014年8月底栖生物总密度分布　Distribution of total benthic density in August 2014 ··················· 383
   2014年11月底栖生物总密度分布　Distribution of total benthic density in November 2014 ··············· 383
   2015年2月底栖生物总密度分布　Distribution of total benthic density in February 2015 ··············· 384
   2015年5月底栖生物总密度分布　Distribution of total benthic density in May 2015 ······················ 384
   2015年8月底栖生物总密度分布　Distribution of total benthic density in August 2015 ··················· 385
   2015年11月底栖生物总密度分布　Distribution of total benthic density in November 2015 ··············· 385

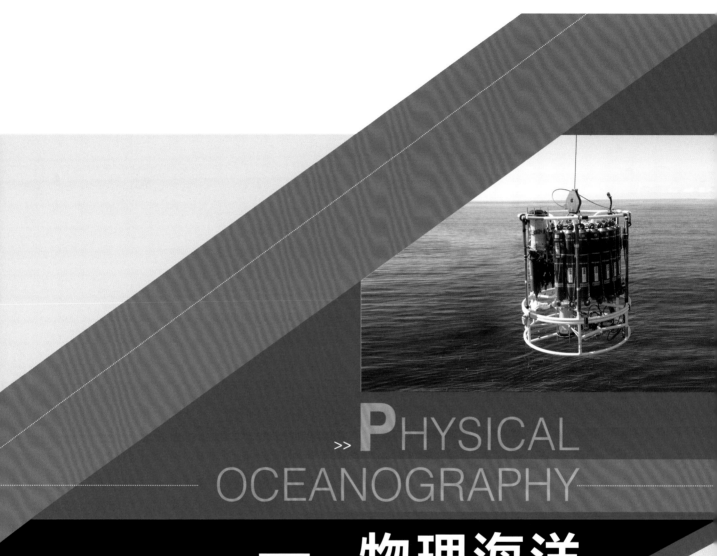

一、物理海洋

PHYSICAL OCEANOGRAPHY

1. 温度 Temperature

2010年2月温度分布
Distribution of temperature in February 2010

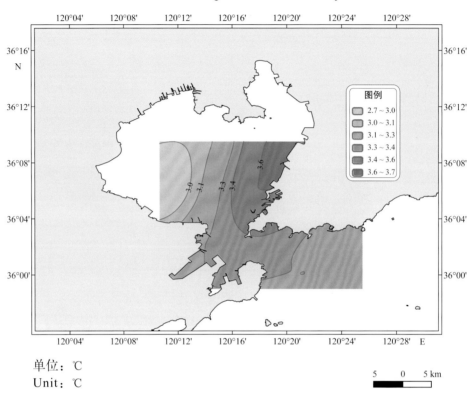

单位：℃
Unit：℃

2010年5月温度分布
Distribution of temperature in May 2010

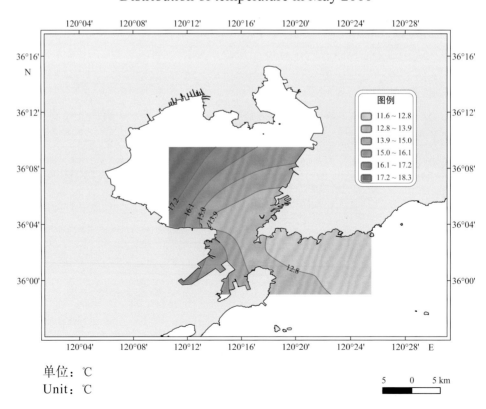

单位：℃
Unit：℃

2010年8月温度分布
Distribution of temperature in August 2010

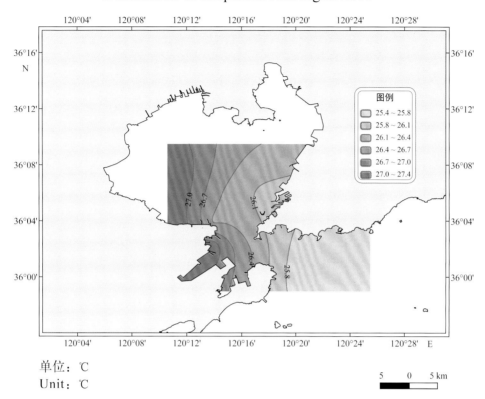

单位：℃
Unit：℃

2010年11月温度分布
Distribution of temperature in November 2010

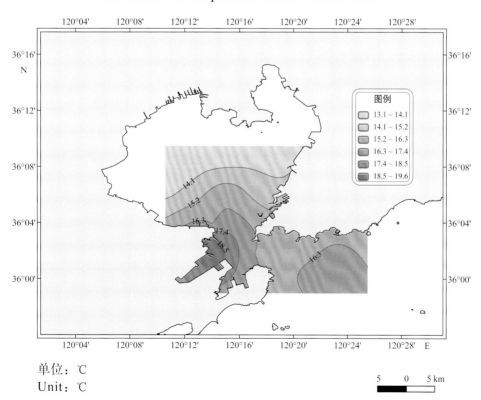

单位：℃
Unit：℃

2011年2月温度分布
Distribution of temperature in February 2011

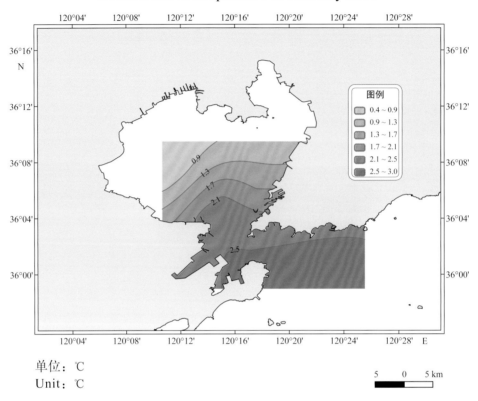

单位：℃
Unit：℃

2011年5月温度分布
Distribution of temperature in May 2011

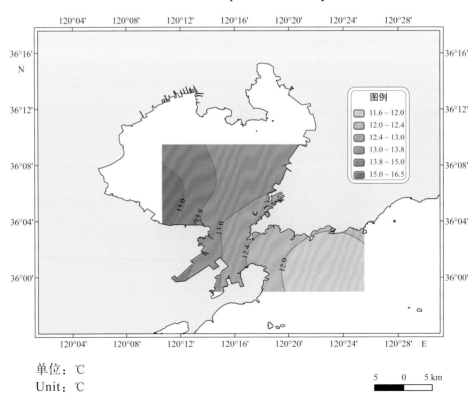

单位：℃
Unit：℃

2011年8月温度分布
Distribution of temperature in August 2011

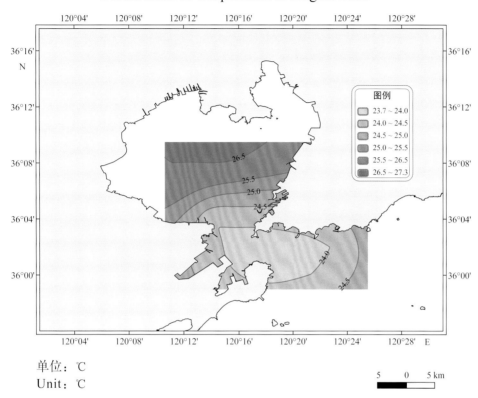

单位：℃
Unit：℃

2011年11月温度分布
Distribution of temperature in November 2011

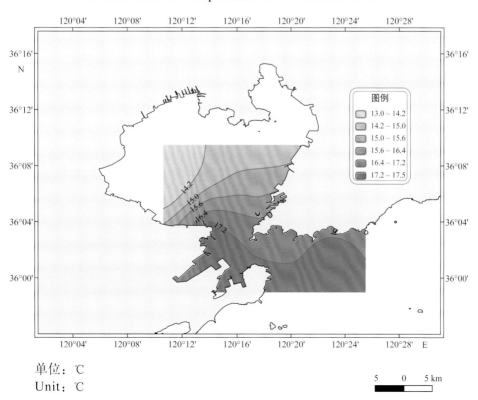

单位：℃
Unit：℃

2012年2月温度分布
Distribution of temperature in February 2012

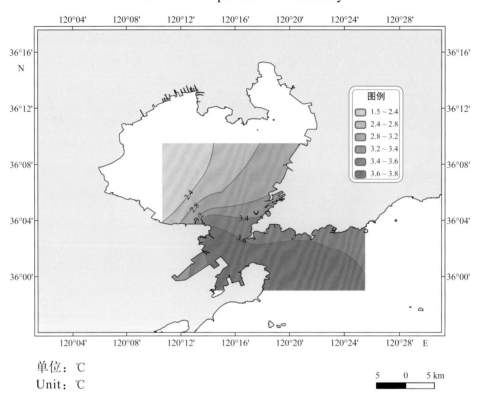

单位：℃
Unit：℃

2012年5月温度分布
Distribution of temperature in May 2012

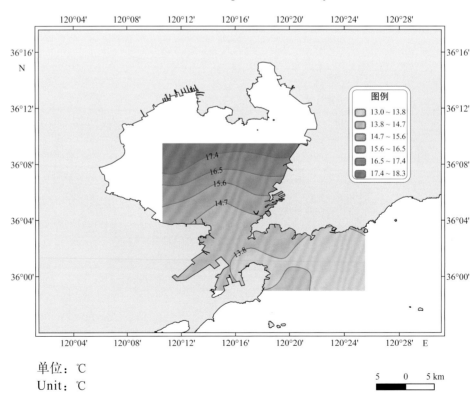

单位：℃
Unit：℃

2012年8月温度分布
Distribution of temperature in August 2012

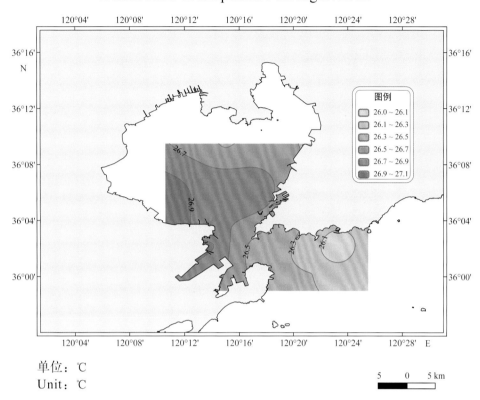

单位：℃
Unit：℃

2012年11月温度分布
Distribution of temperature in November 2012

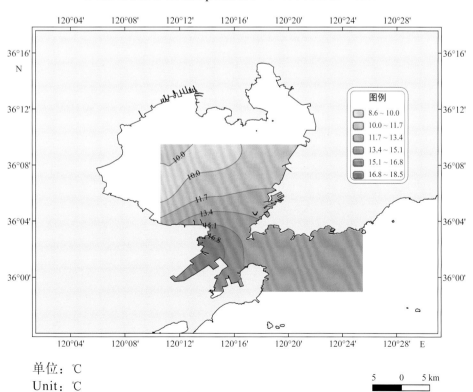

单位：℃
Unit：℃

2013年2月温度分布
Distribution of temperature in February 2013

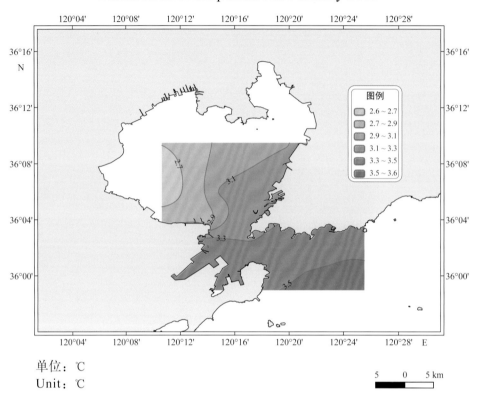

单位：℃
Unit：℃

2013年5月温度分布
Distribution of temperature in May 2013

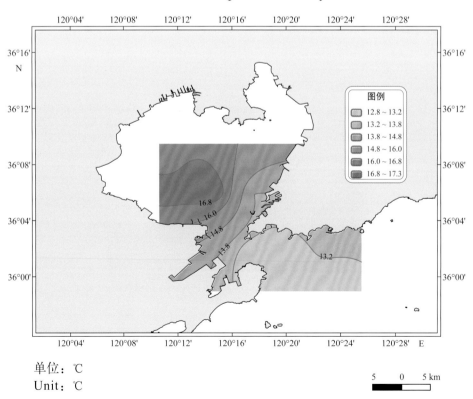

单位：℃
Unit：℃

2013年8月温度分布
Distribution of temperature in August 2013

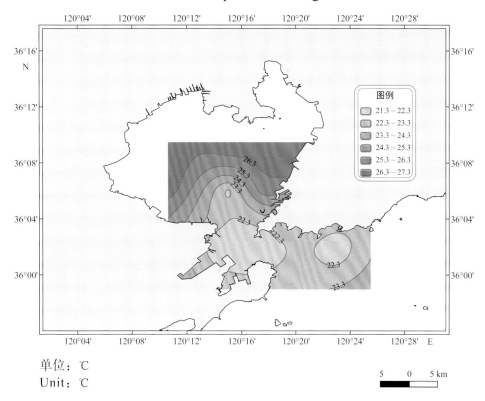

单位：℃
Unit：℃

2013年11月温度分布
Distribution of temperature in November 2013

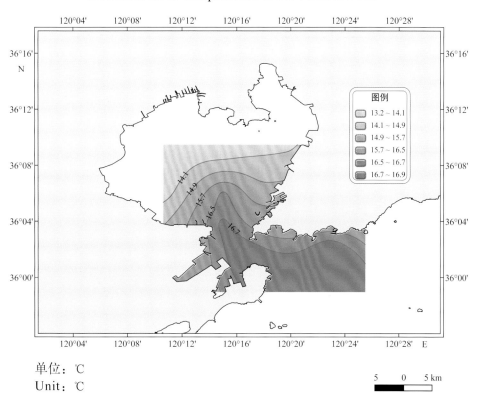

单位：℃
Unit：℃

2014年2月温度分布
Distribution of temperature in February 2014

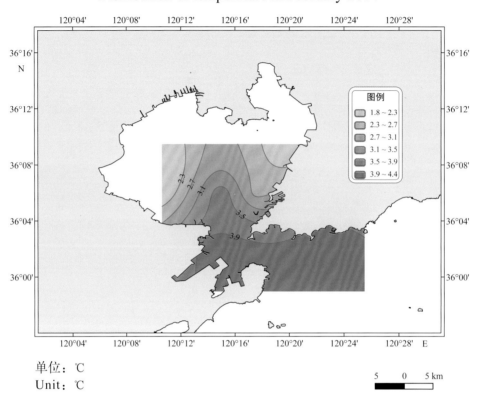

单位：℃
Unit：℃

2014年5月温度分布
Distribution of temperature in May 2014

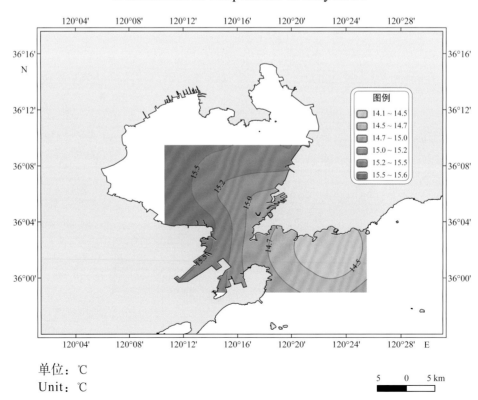

单位：℃
Unit：℃

2014年8月温度分布
Distribution of temperature in August 2014

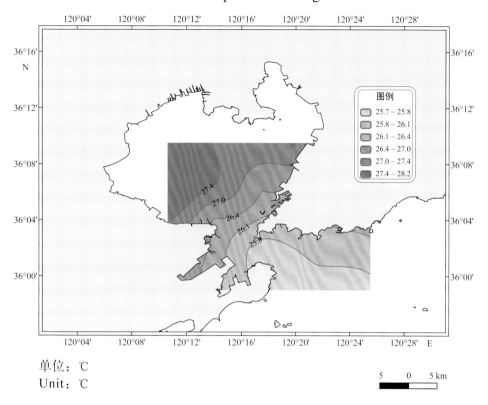

单位：℃
Unit：℃

2014年11月温度分布
Distribution of temperature in November 2014

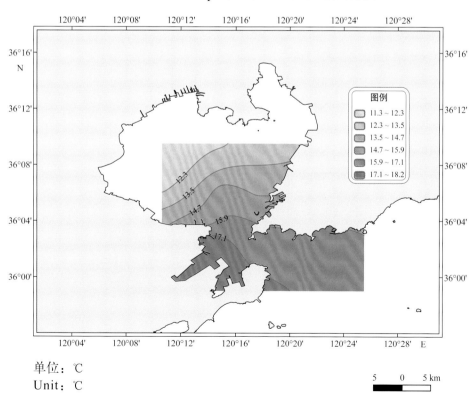

单位：℃
Unit：℃

2015年2月温度分布
Distribution of temperature in February 2015

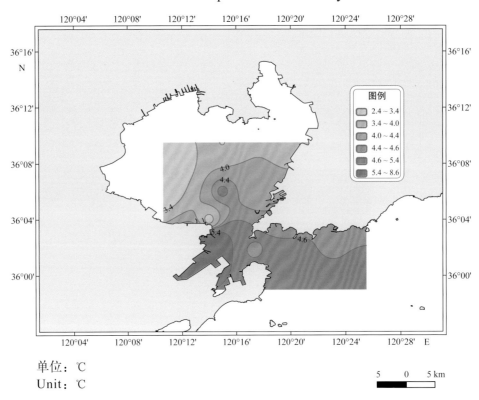

单位：℃
Unit：℃

2015年5月温度分布
Distribution of temperature in May 2015

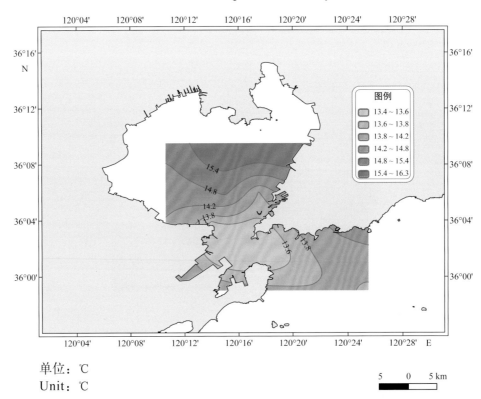

单位：℃
Unit：℃

2015年8月温度分布
Distribution of temperature in August 2015

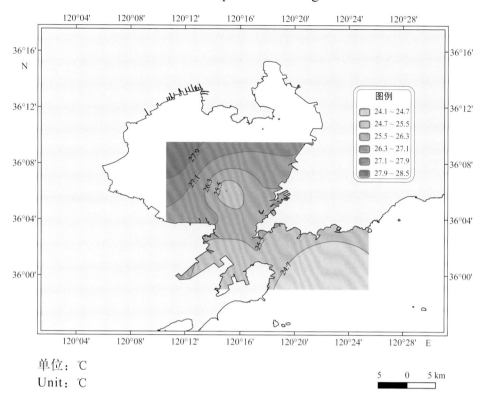

单位：℃
Unit：℃

2015年11月温度分布
Distribution of temperature in November 2015

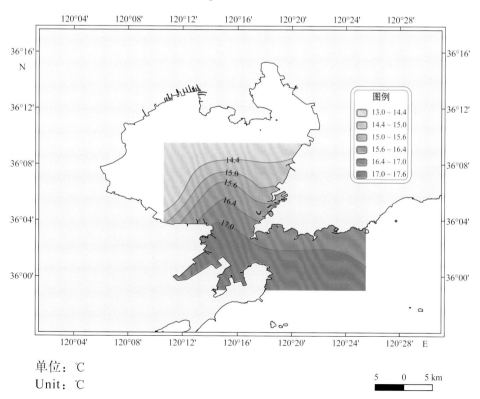

单位：℃
Unit：℃

2. 盐度 Salinity

2010年2月盐度分布
Distribution of salinity in February 2010

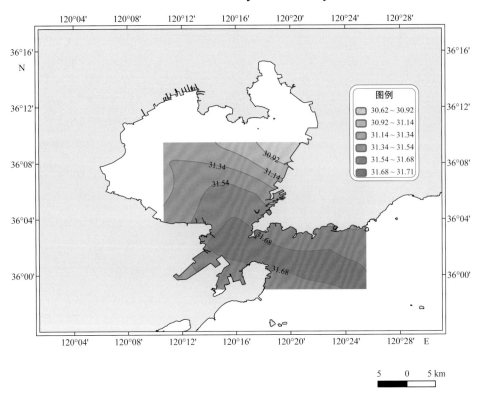

2010年5月盐度分布
Distribution of salinity in May 2010

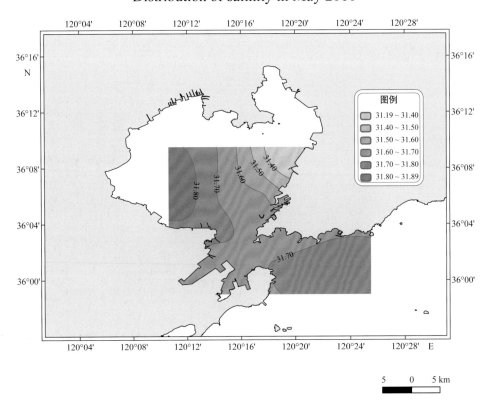

2010年8月盐度分布
Distribution of salinity in August 2010

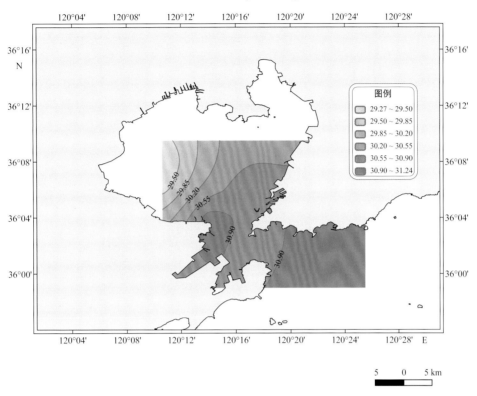

2010年11月盐度分布
Distribution of salinity in November 2010

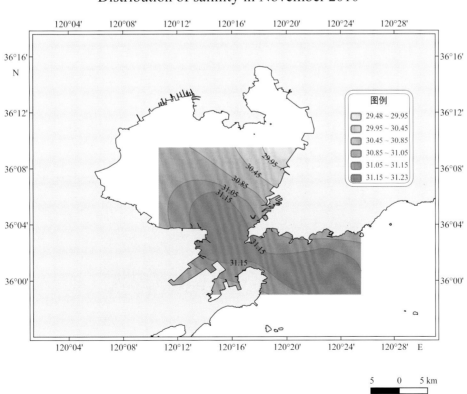

2011年2月盐度分布
Distribution of salinity in February 2011

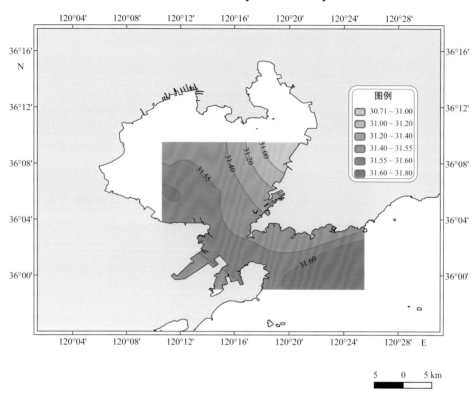

2011年5月盐度分布
Distribution of salinity in May 2011

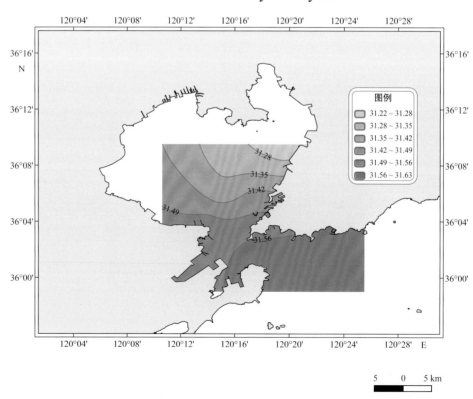

2011年8月盐度分布
Distribution of salinity in August 2011

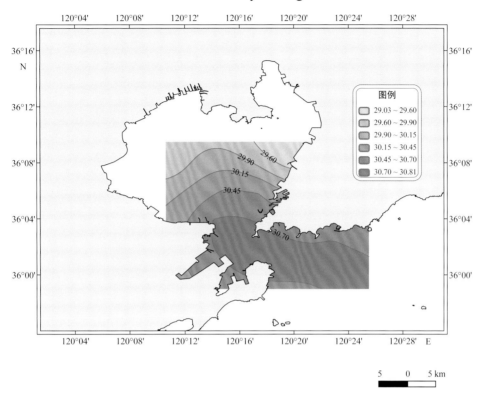

2011年11月盐度分布
Distribution of salinity in November 2011

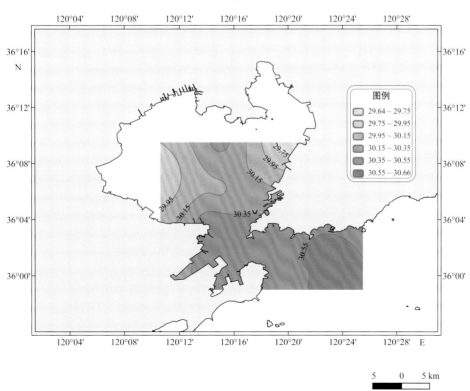

2012年2月盐度分布
Distribution of salinity in February 2012

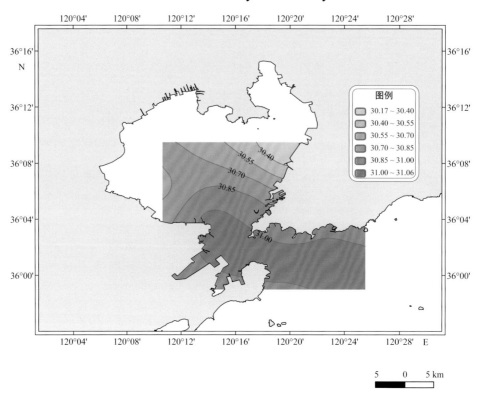

2012年5月盐度分布
Distribution of salinity in May 2012

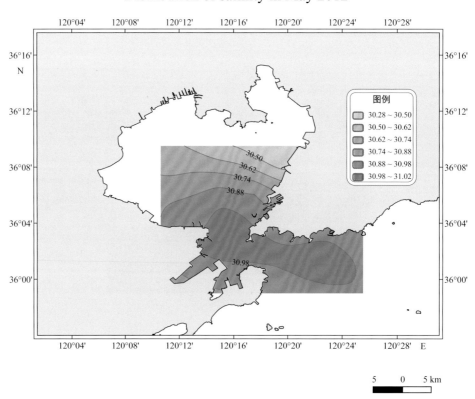

2012年8月盐度分布
Distribution of salinity in August 2012

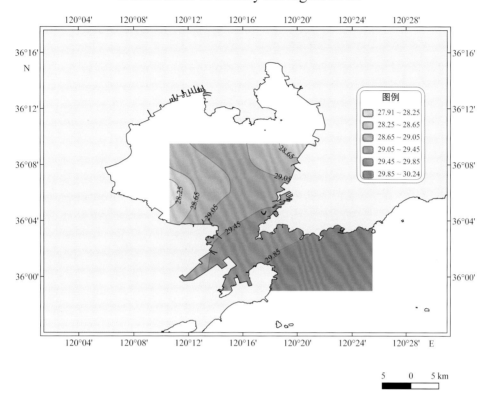

2012年11月盐度分布
Distribution of salinity in November 2012

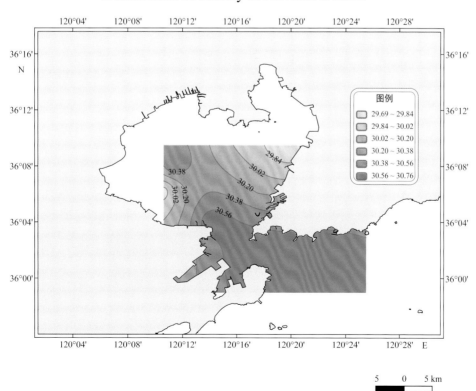

2013年2月盐度分布
Distribution of salinity in February 2013

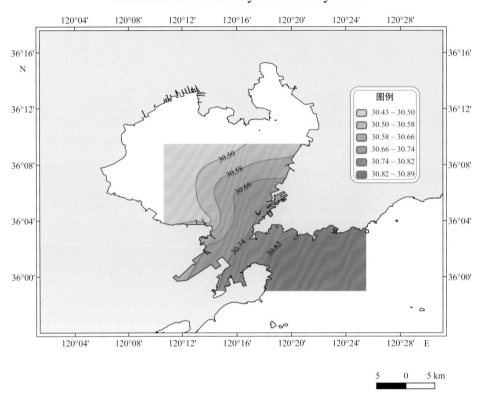

2013年5月盐度分布
Distribution of salinity in May 2013

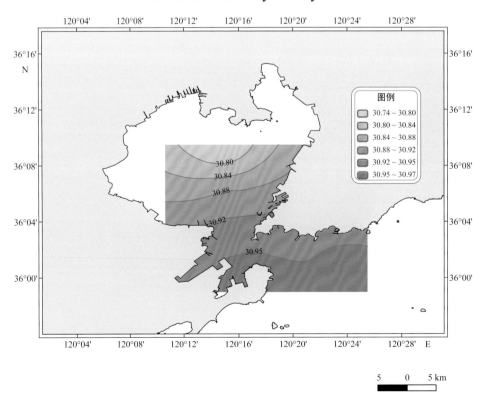

2013年8月盐度分布
Distribution of salinity in August 2013

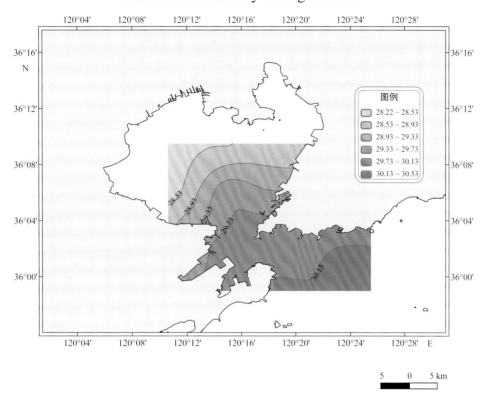

2013年11月盐度分布
Distribution of salinity in November 2013

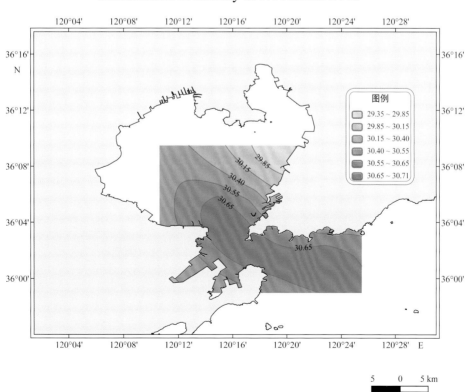

2014年2月盐度分布
Distribution of salinity in February 2014

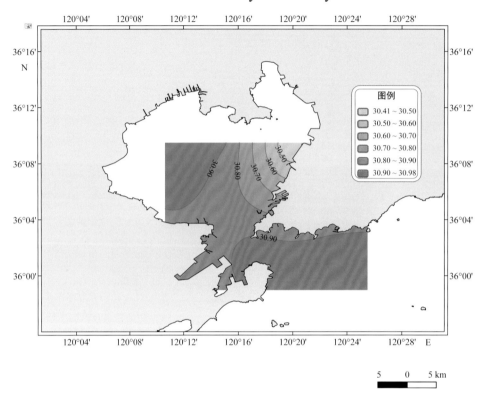

2014年5月盐度分布
Distribution of salinity in May 2014

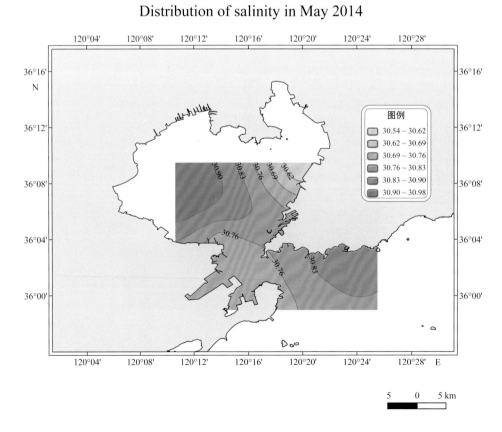

2014年8月盐度分布
Distribution of salinity in August 2014

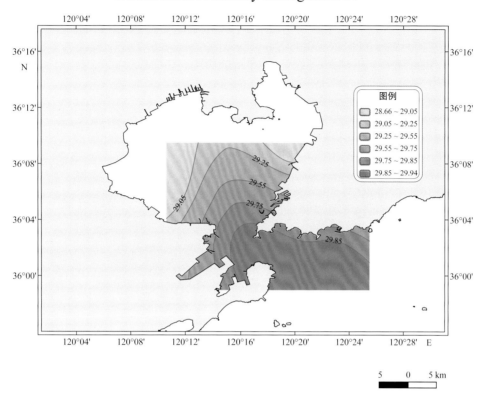

2014年11月盐度分布
Distribution of salinity in November 2014

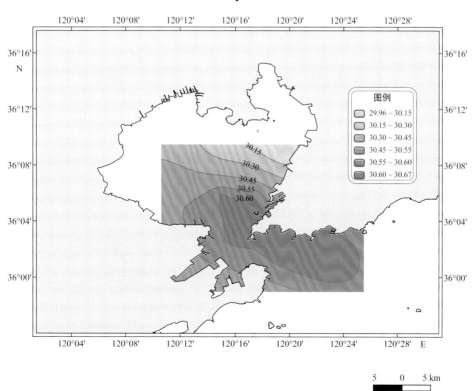

2015年2月盐度分布
Distribution of salinity in February 2015

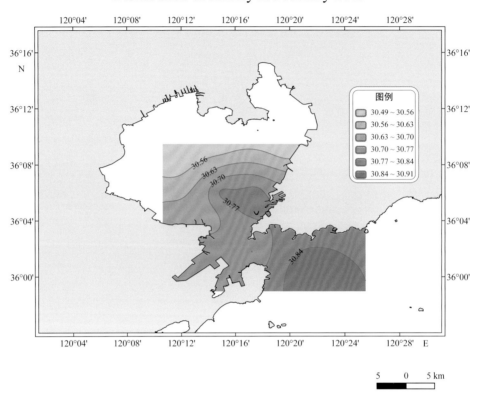

2015年5月盐度分布
Distribution of salinity in May 2015

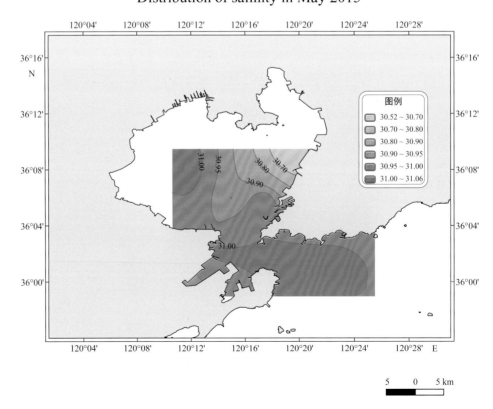

2015年8月盐度分布
Distribution of salinity in August 2015

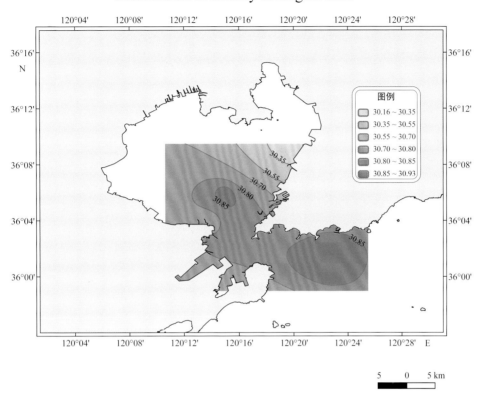

2015年11月盐度分布
Distribution of salinity in November 2015

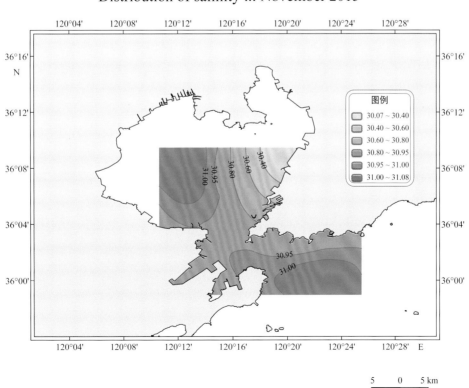

3. 透明度 Secchi disk depth

2010年2月透明度分布
Distribution of secchi disk depth in February 2010

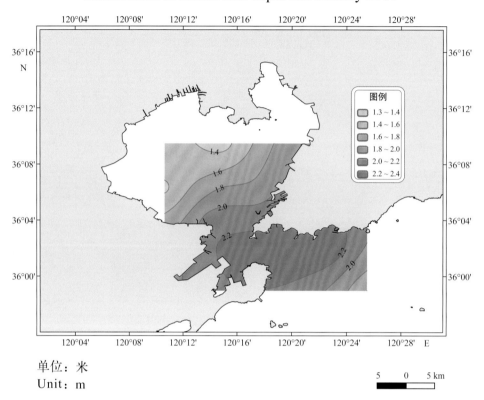

单位：米
Unit：m

2010年5月透明度分布
Distribution of secchi disk depth in May 2010

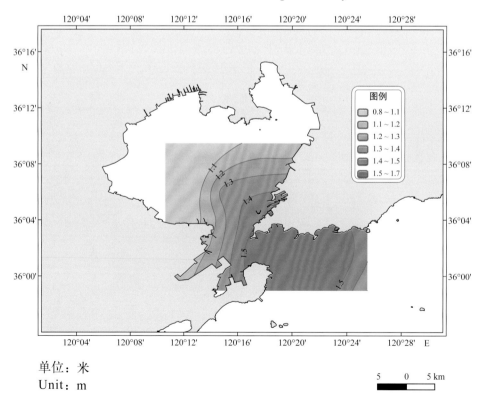

单位：米
Unit：m

2010年8月透明度分布
Distribution of secchi disk depth in August 2010

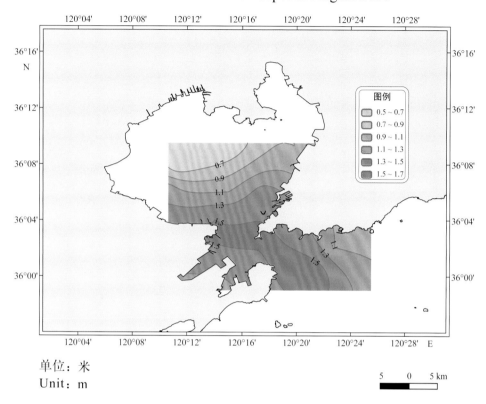

单位：米
Unit：m

2010年11月透明度分布
Distribution of secchi disk depth in November 2010

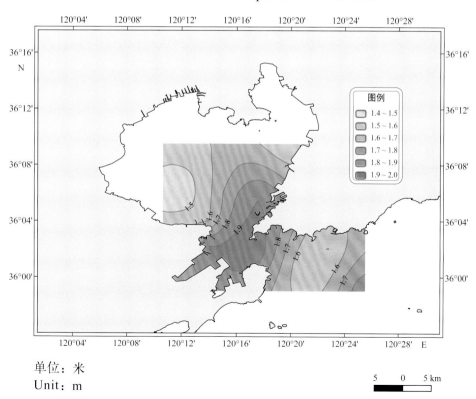

单位：米
Unit：m

2011年2月透明度分布
Distribution of secchi disk depth in February 2011

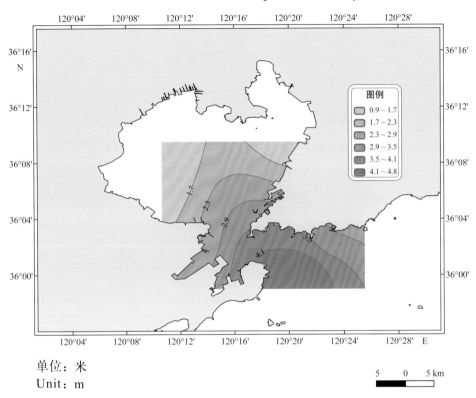

单位：米
Unit：m

2011年5月透明度分布
Distribution of secchi disk depth in May 2011

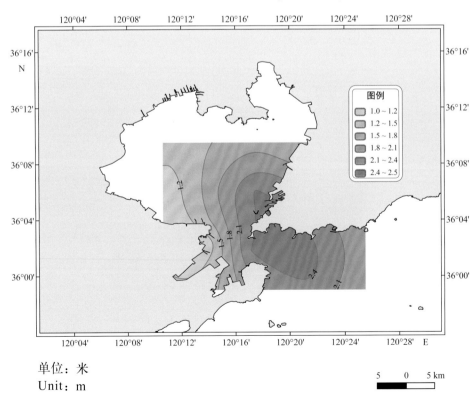

单位：米
Unit：m

2011年8月透明度分布
Distribution of secchi disk depth in August 2011

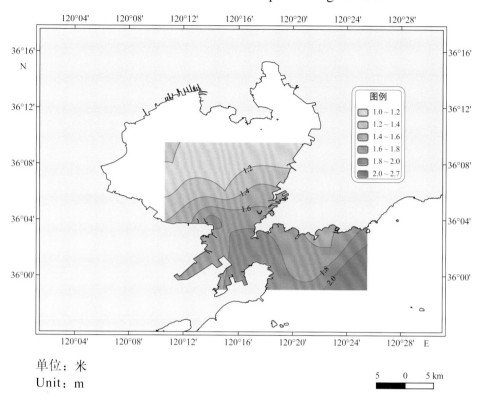

单位：米
Unit：m

2011年11月透明度分布
Distribution of secchi disk depth in November 2011

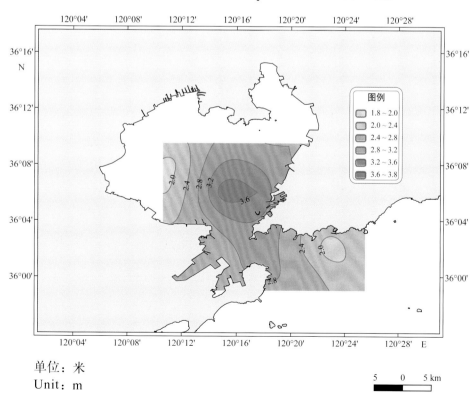

单位：米
Unit：m

2012年2月透明度分布
Distribution of secchi disk depth in February 2012

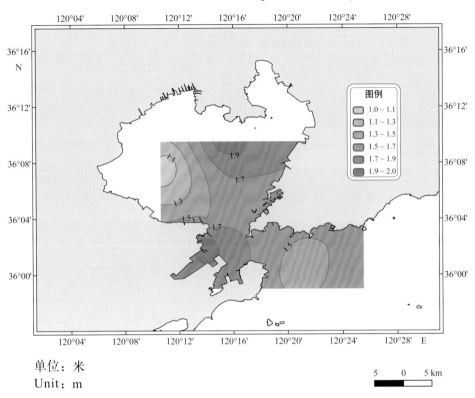

单位：米
Unit：m

2012年5月透明度分布
Distribution of secchi disk depth in May 2012

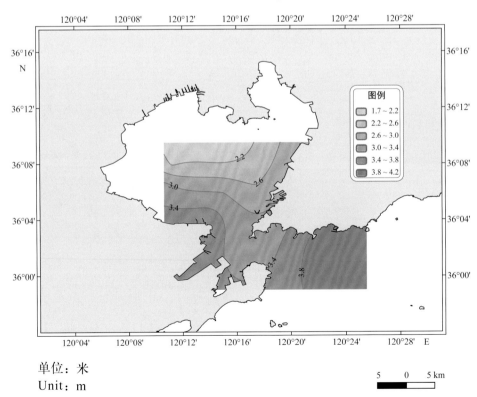

单位：米
Unit：m

2012年8月透明度分布
Distribution of secchi disk depth in August 2012

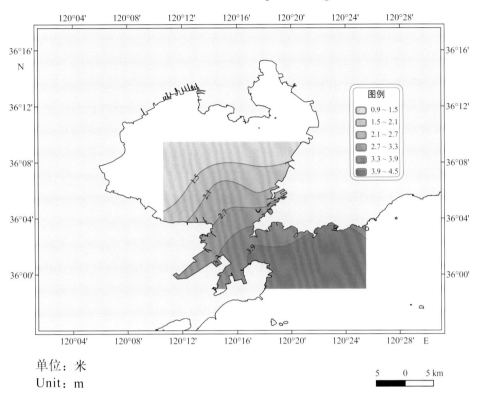

单位：米
Unit：m

2012年11月透明度分布
Distribution of secchi disk depth in November 2012

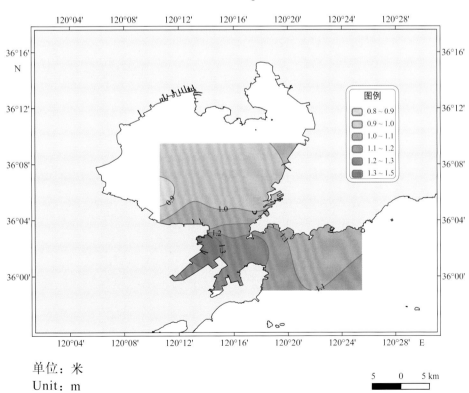

单位：米
Unit：m

2013年2月透明度分布
Distribution of secchi disk depth in February 2013

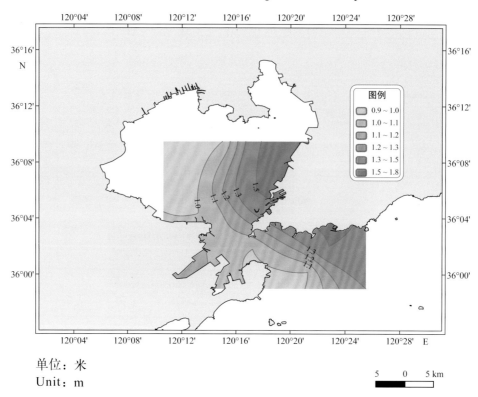

单位：米
Unit：m

2013年5月透明度分布
Distribution of secchi disk depth in May 2013

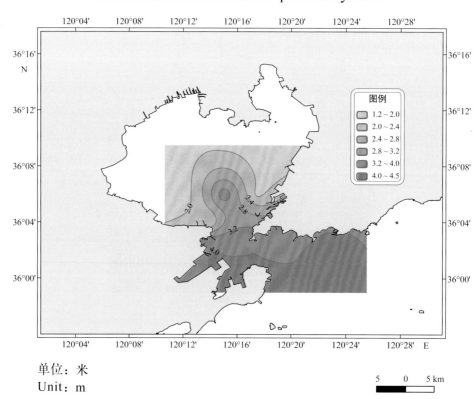

单位：米
Unit：m

2013年8月透明度分布
Distribution of secchi disk depth in August 2013

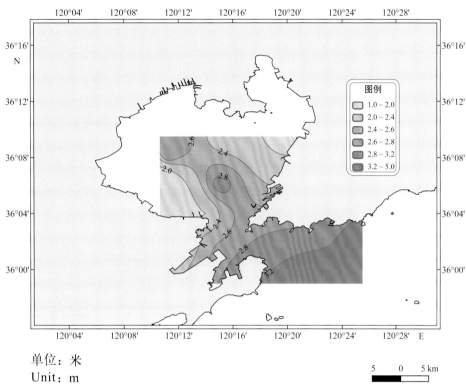

单位：米
Unit：m

2013年11月透明度分布
Distribution of secchi disk depth in November 2013

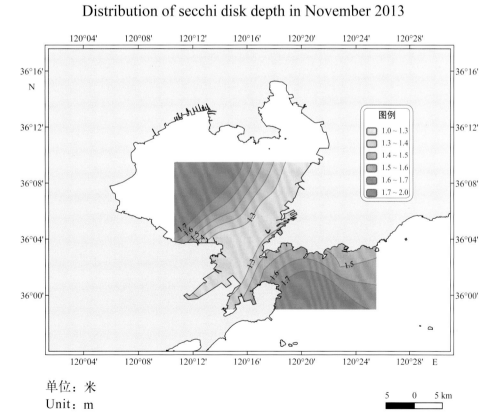

单位：米
Unit：m

2014年2月透明度分布
Distribution of secchi disk depth in February 2014

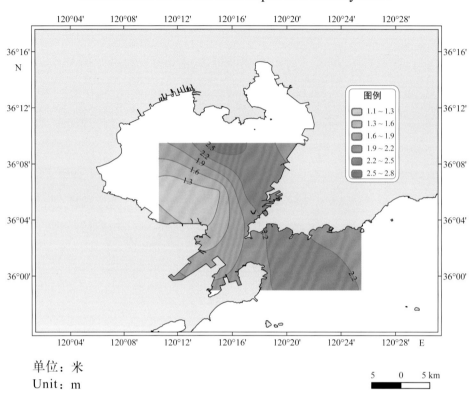

单位：米
Unit：m

2014年5月透明度分布
Distribution of secchi disk depth in May 2014

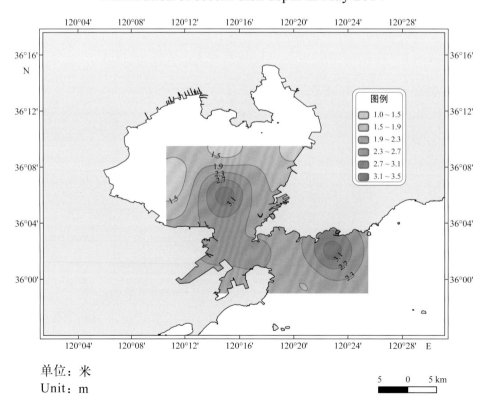

单位：米
Unit：m

2014年8月透明度分布
Distribution of secchi disk depth in August 2014

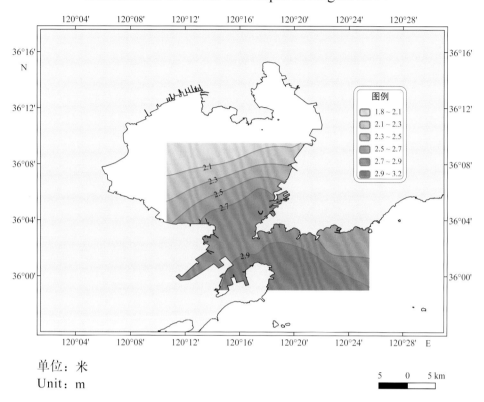

单位：米
Unit：m

2014年11月透明度分布
Distribution of secchi disk depth in November 2014

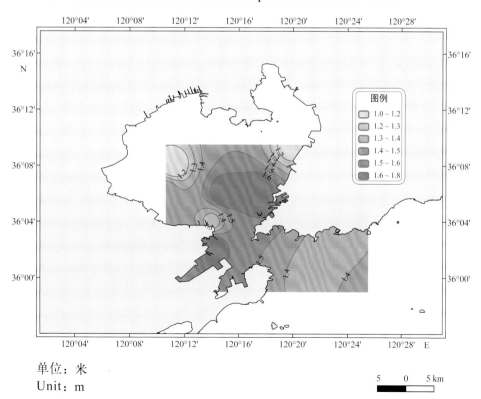

单位：米
Unit：m

2015年2月透明度分布
Distribution of secchi disk depth in February 2015

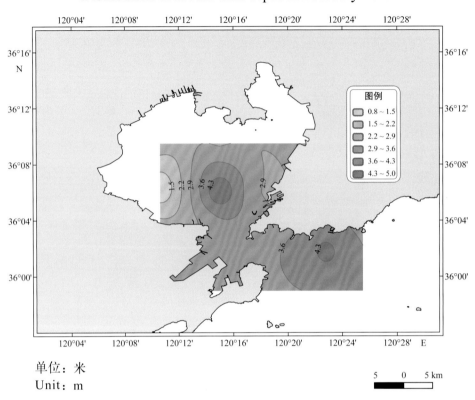

单位：米
Unit：m

2015年5月透明度分布
Distribution of secchi disk depth in May 2015

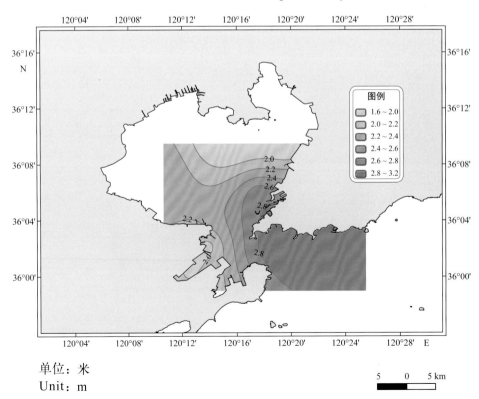

单位：米
Unit：m

2015年8月透明度分布
Distribution of secchi disk depth in August 2015

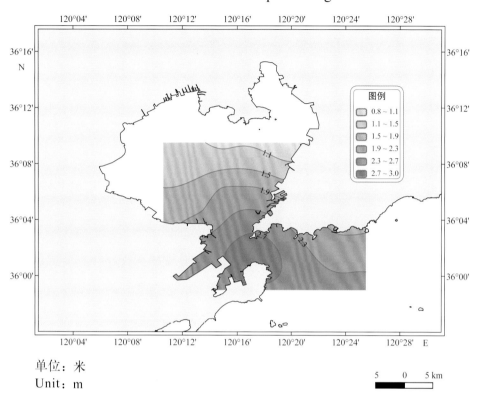

单位：米
Unit：m

2015年11月透明度分布
Distribution of secchi disk depth in November 2015

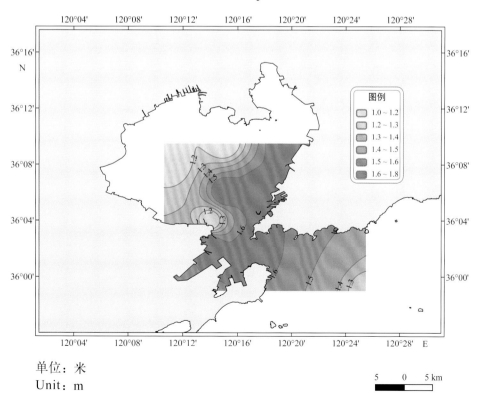

单位：米
Unit：m

CHEMICAL OCEANOGRAPHY

二、化学海洋

1. pH 值 pH

2010年2月pH值分布
Distribution of pH in February 2010

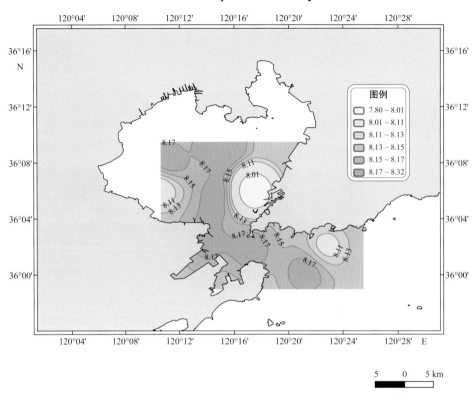

2010年5月pH值分布
Distribution of pH in May 2010

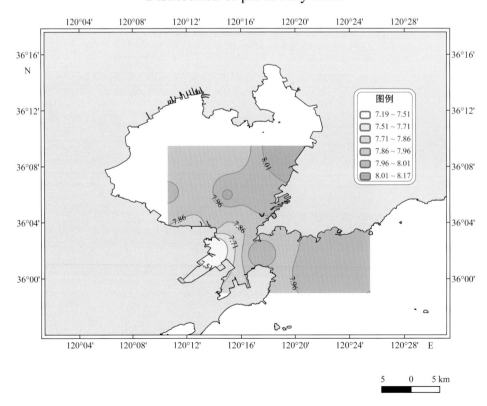

2010年8月pH值分布
Distribution of pH in August 2010

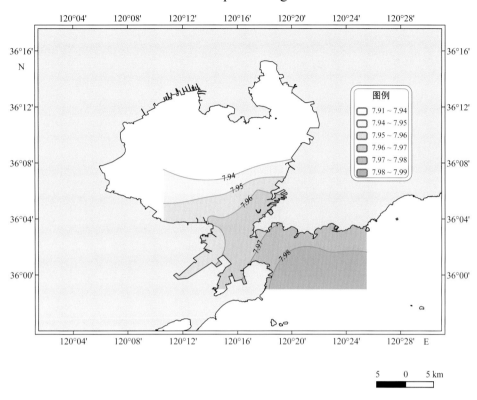

2010年11月pH值分布
Distribution of pH in November 2010

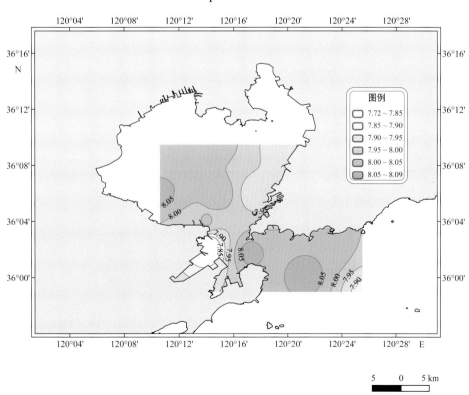

2011年2月pH值分布
Distribution of pH in February 2011

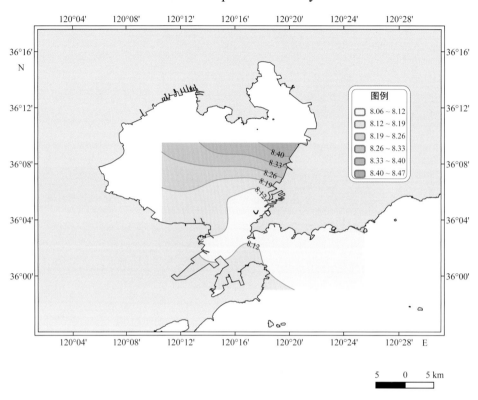

2011年5月pH值分布
Distribution of pH in May 2011

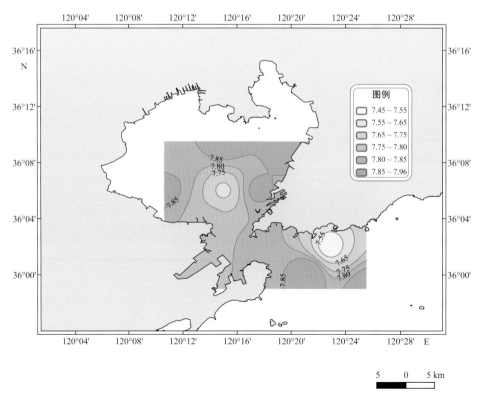

2011年8月pH值分布
Distribution of pH in August 2011

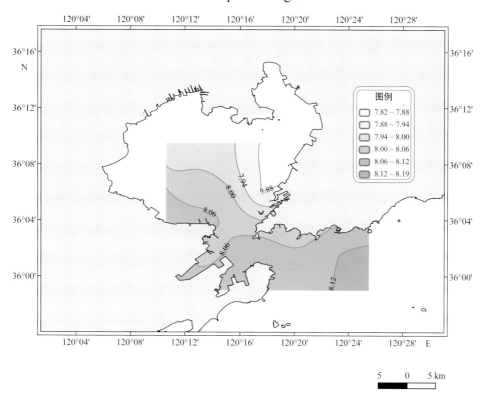

2011年11月pH值分布
Distribution of pH in November 2011

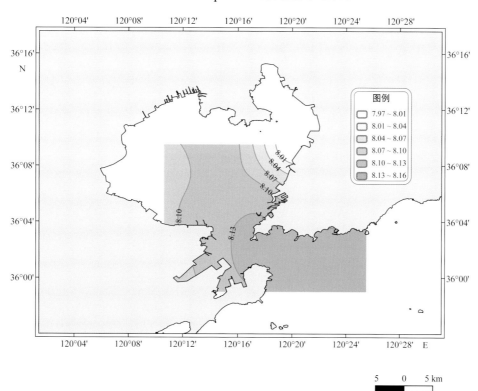

2012年2月pH值分布
Distribution of pH in February 2012

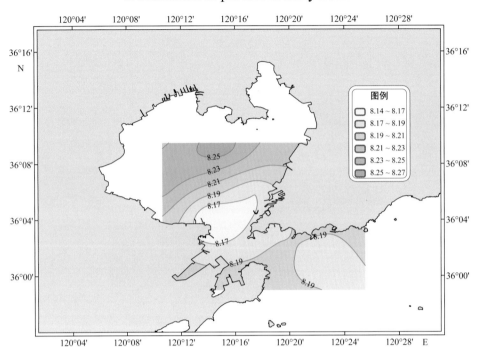

2012年5月pH值分布
Distribution of pH in May 2012

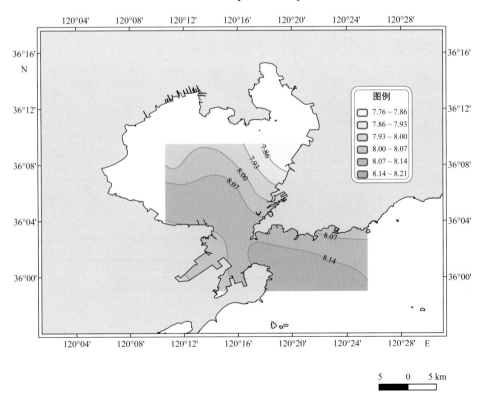

2012年8月pH值分布
Distribution of pH in August 2012

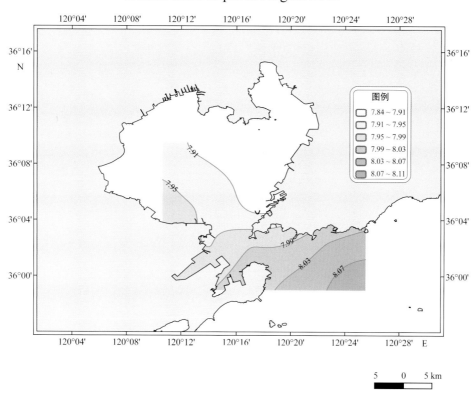

2012年11月pH值分布
Distribution of pH in November 2012

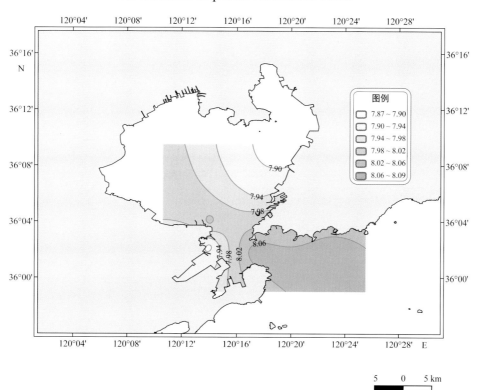

2013年2月pH值分布
Distribution of pH in February 2013

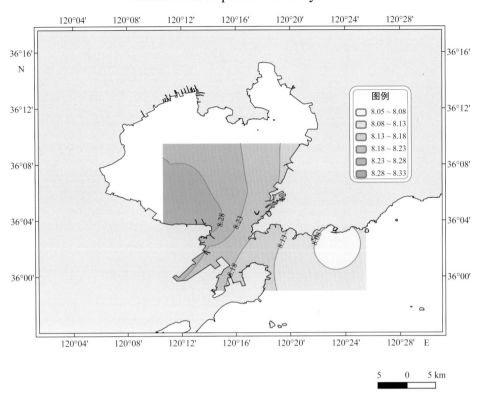

2013年5月pH值分布
Distribution of pH in May 2013

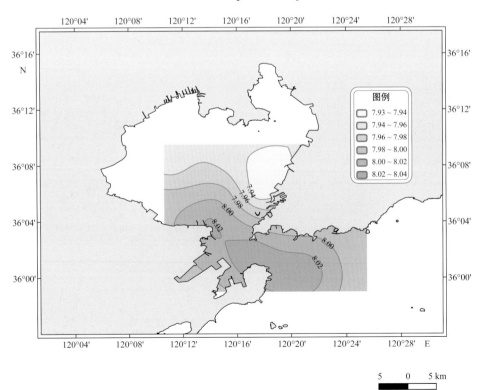

2013年8月pH值分布
Distribution of pH in August 2013

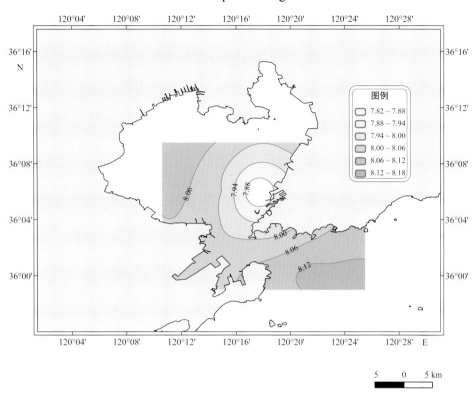

2013年11月pH值分布
Distribution of pH in November 2013

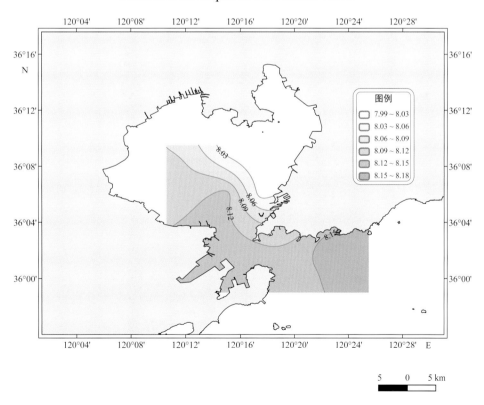

2014年2月pH值分布
Distribution of pH in February 2014

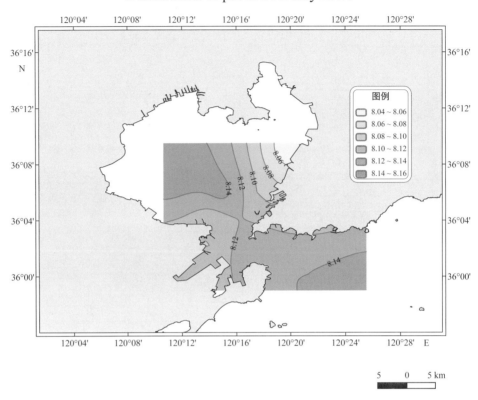

2014年5月pH值分布
Distribution of pH in May 2014

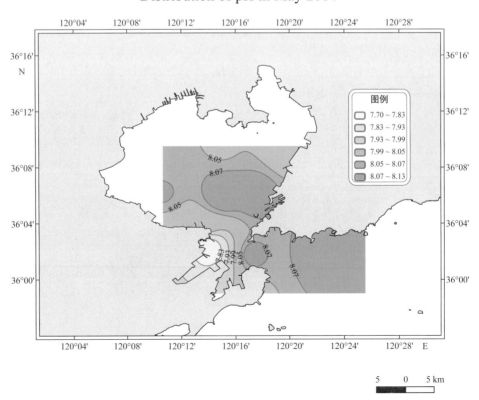

2014年8月pH值分布
Distribution of pH in August 2014

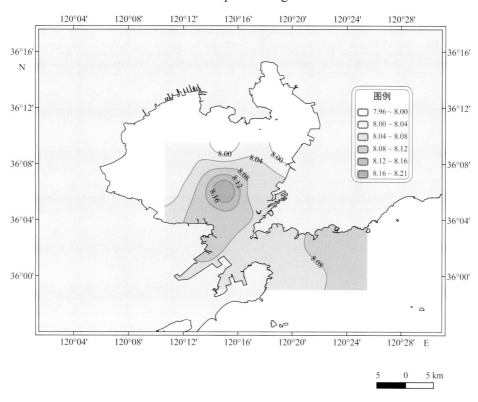

2014年11月pH值分布
Distribution of pH in November 2014

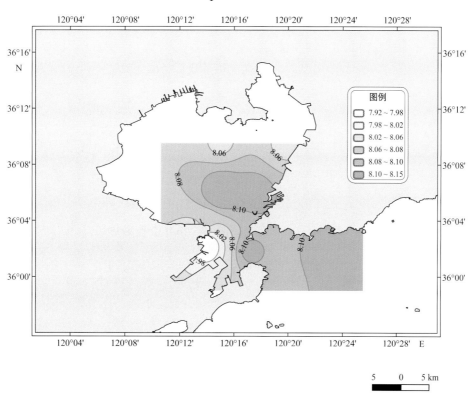

2015年2月pH值分布
Distribution of pH in February 2015

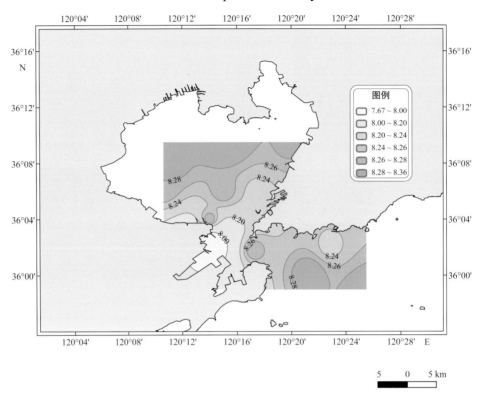

2015年5月pH值分布
Distribution of pH in May 2015

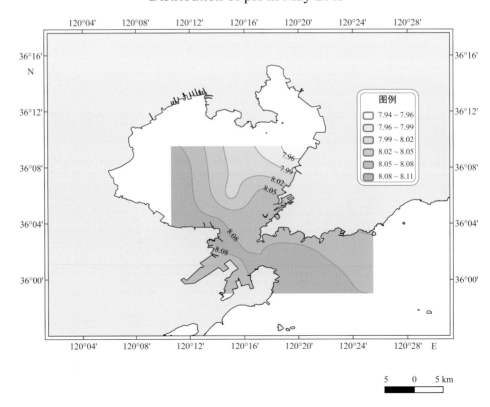

2015年8月pH值分布
Distribution of pH in August 2015

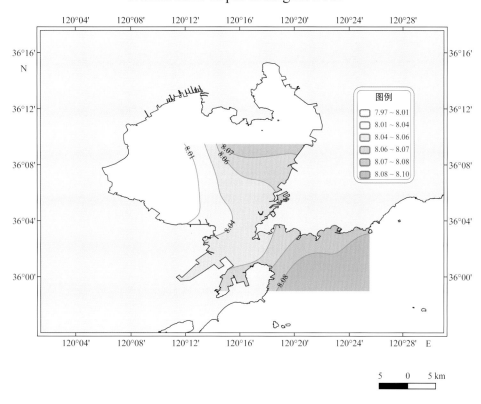

2015年11月pH值分布
Distribution of pH in November 2015

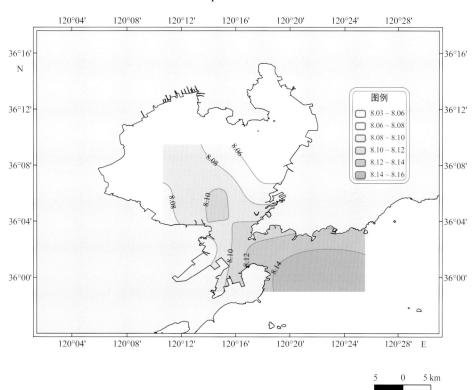

2. 溶解氧 Dissolved oxygen

2010年2月溶解氧分布
Distribution of dissolved oxygen in February 2010

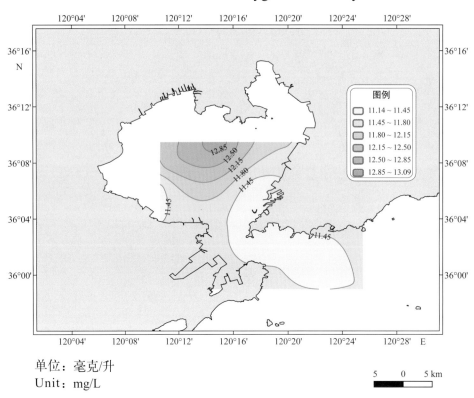

单位：毫克/升
Unit：mg/L

2010年5月溶解氧分布
Distribution of dissolved oxygen in May 2010

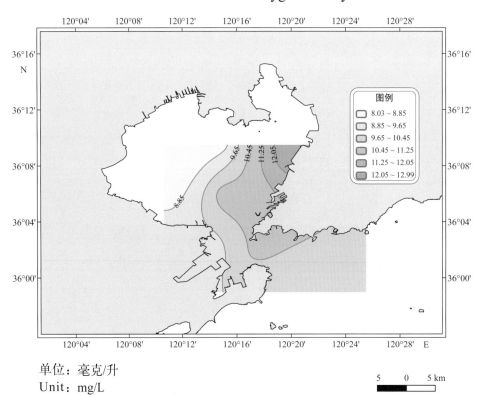

单位：毫克/升
Unit：mg/L

2010年8月溶解氧分布
Distribution of dissolved oxygen in August 2010

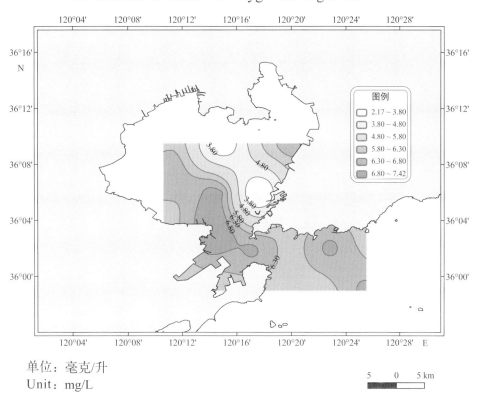

单位：毫克/升
Unit：mg/L

2010年11月溶解氧分布
Distribution of dissolved oxygen in November 2010

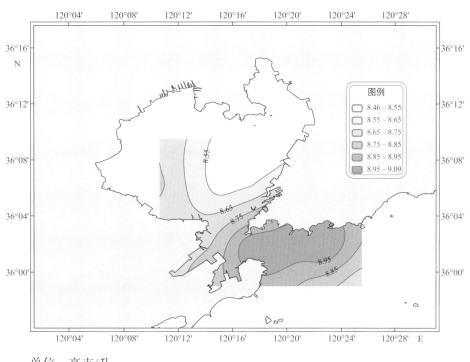

单位：毫克/升
Unit：mg/L

2011年2月溶解氧分布
Distribution of dissolved oxygen in February 2011

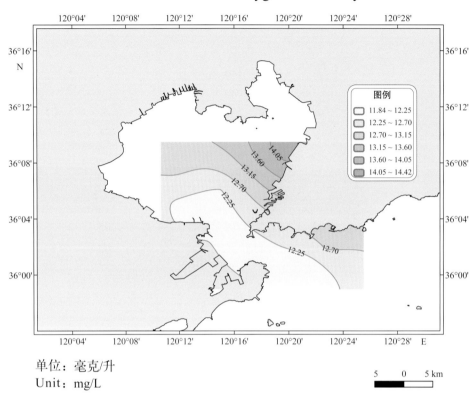

单位：毫克/升
Unit：mg/L

2011年5月溶解氧分布
Distribution of dissolved oxygen in May 2011

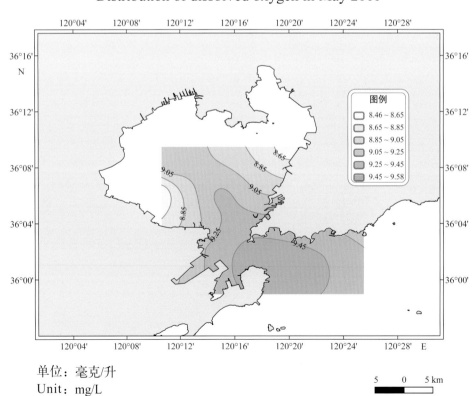

单位：毫克/升
Unit：mg/L

2011年8月溶解氧分布
Distribution of dissolved oxygen in August 2011

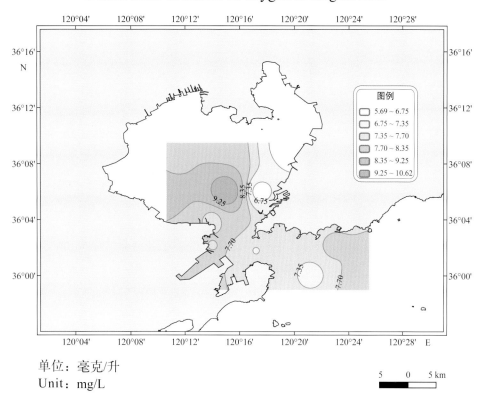

单位：毫克/升
Unit：mg/L

2011年11月溶解氧分布
Distribution of dissolved oxygen in November 2011

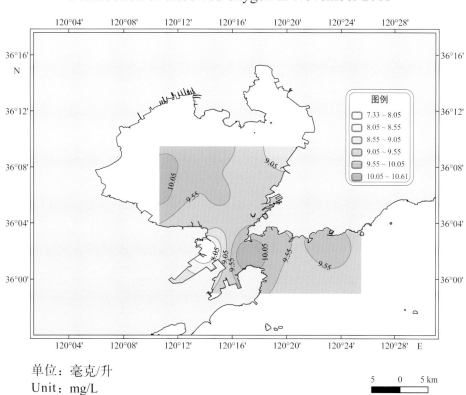

单位：毫克/升
Unit：mg/L

2012年2月溶解氧分布
Distribution of dissolved oxygen in February 2012

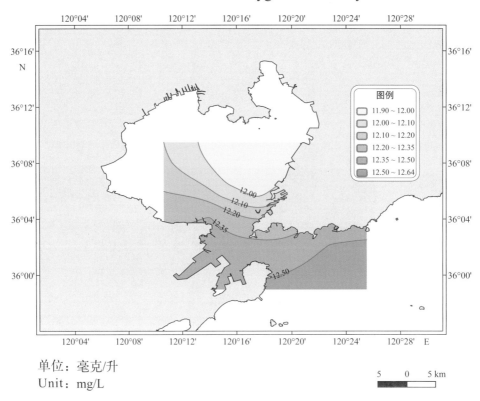

单位：毫克/升
Unit: mg/L

2012年5月溶解氧分布
Distribution of dissolved oxygen in May 2012

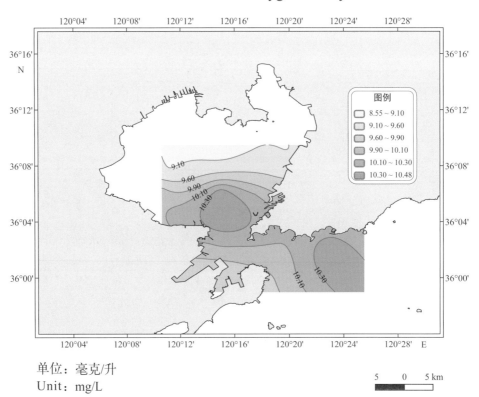

单位：毫克/升
Unit: mg/L

2012年8月溶解氧分布
Distribution of dissolved oxygen in August 2012

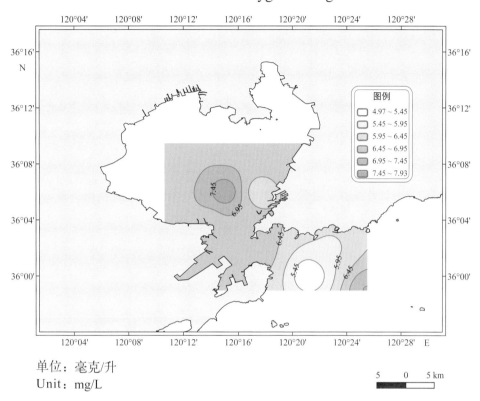

单位：毫克/升
Unit：mg/L

2012年11月溶解氧分布
Distribution of dissolved oxygen in November 2012

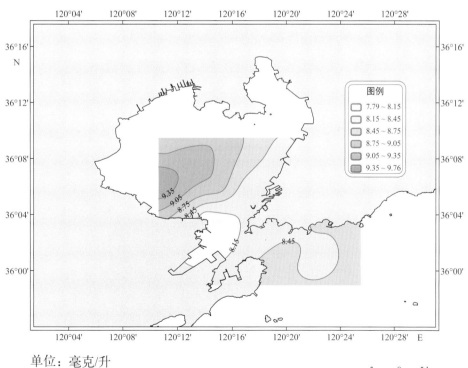

单位：毫克/升
Unit：mg/L

2013年2月溶解氧分布
Distribution of dissolved oxygen in February 2013

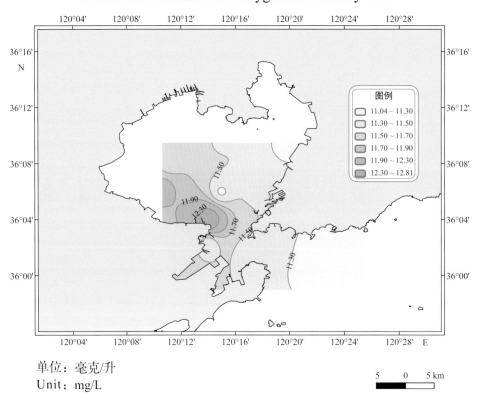

单位：毫克/升
Unit：mg/L

2013年5月溶解氧分布
Distribution of dissolved oxygen in May 2013

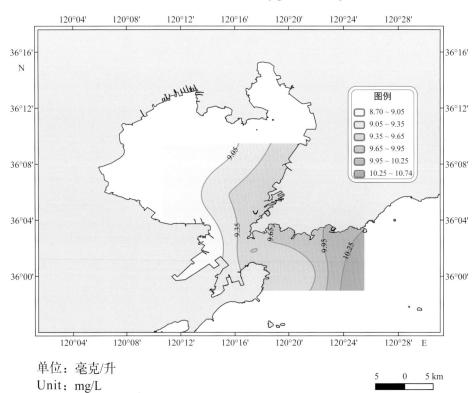

单位：毫克/升
Unit：mg/L

2013年8月溶解氧分布
Distribution of dissolved oxygen in August 2013

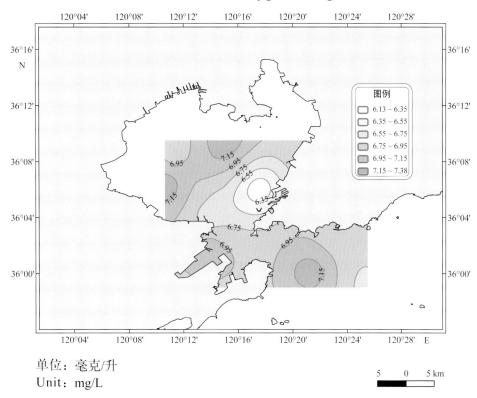

单位：毫克/升
Unit：mg/L

2013年11月溶解氧分布
Distribution of dissolved oxygen in November 2013

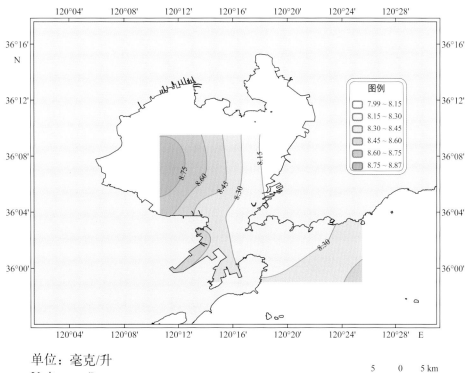

单位：毫克/升
Unit：mg/L

2014年2月溶解氧分布
Distribution of dissolved oxygen in February 2014

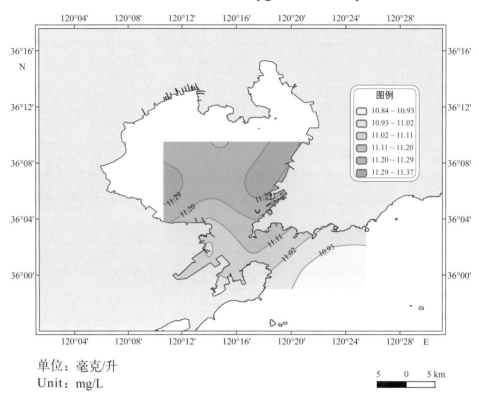

单位：毫克/升
Unit：mg/L

2014年5月溶解氧分布
Distribution of dissolved oxygen in May 2014

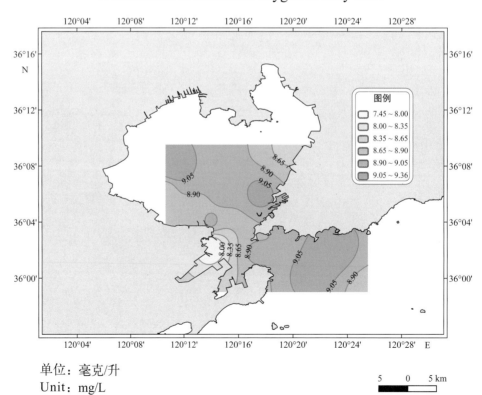

单位：毫克/升
Unit：mg/L

2014年8月溶解氧分布
Distribution of dissolved oxygen in August 2014

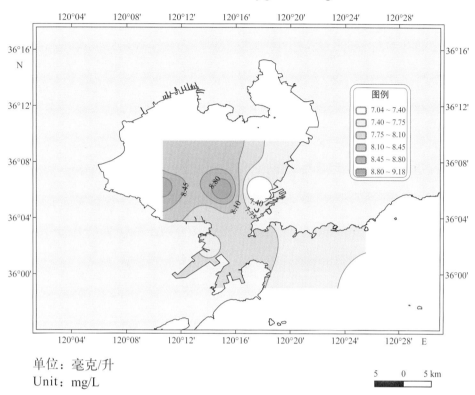

单位：毫克/升
Unit: mg/L

2014年11月溶解氧分布
Distribution of dissolved oxygen in November 2014

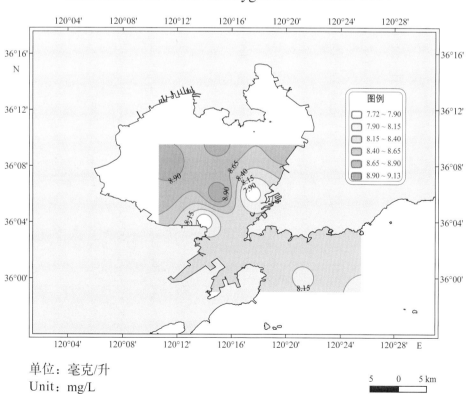

单位：毫克/升
Unit: mg/L

2015年2月溶解氧分布
Distribution of dissolved oxygen in February 2015

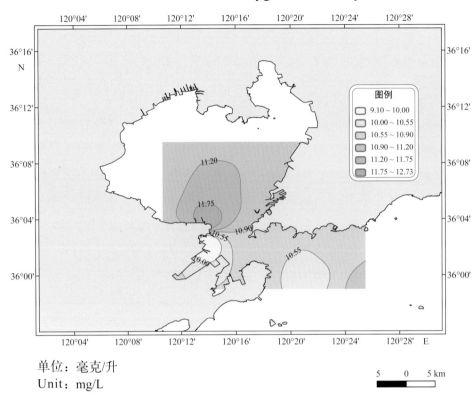

单位：毫克/升
Unit：mg/L

2015年5月溶解氧分布
Distribution of dissolved oxygen in May 2015

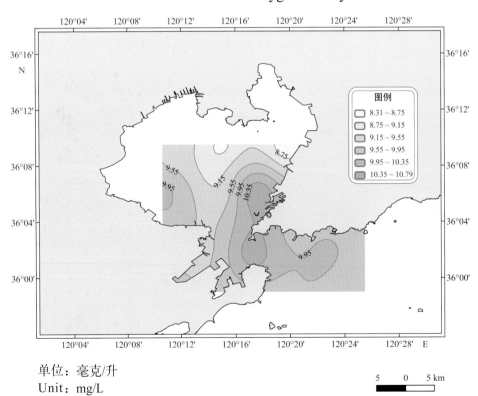

单位：毫克/升
Unit：mg/L

2015年8月溶解氧分布
Distribution of dissolved oxygen in August 2015

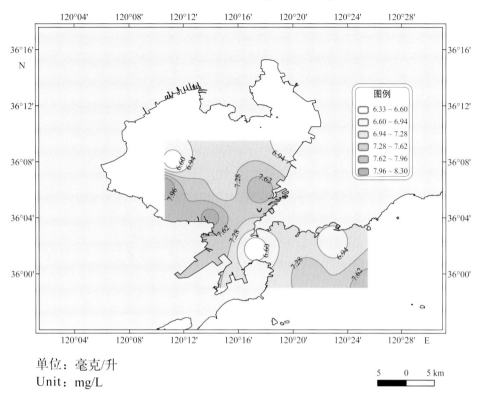

单位：毫克/升
Unit：mg/L

2015年11月溶解氧分布
Distribution of dissolved oxygen in November 2015

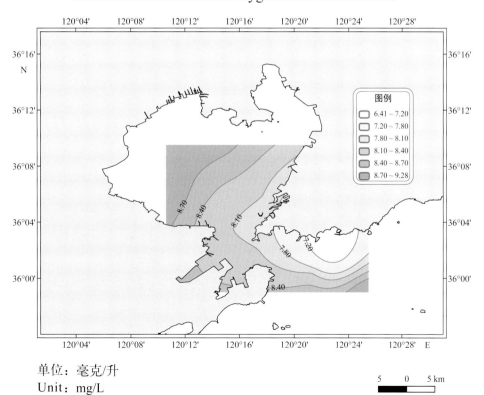

单位：毫克/升
Unit：mg/L

3. 硅酸盐 Silicate

2010年2月硅酸盐分布
Distribution of silicate in February 2010

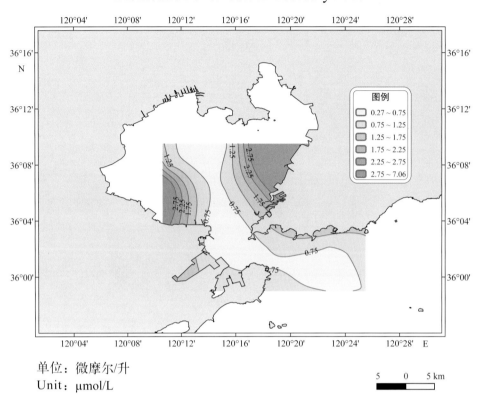

单位：微摩尔/升
Unit：µmol/L

2010年5月硅酸盐分布
Distribution of silicate in May 2010

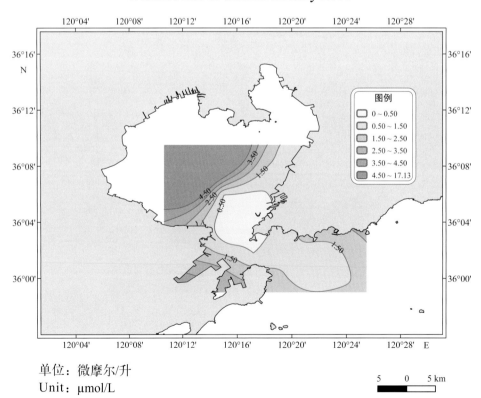

单位：微摩尔/升
Unit：µmol/L

2010年8月硅酸盐分布
Distribution of silicate in August 2010

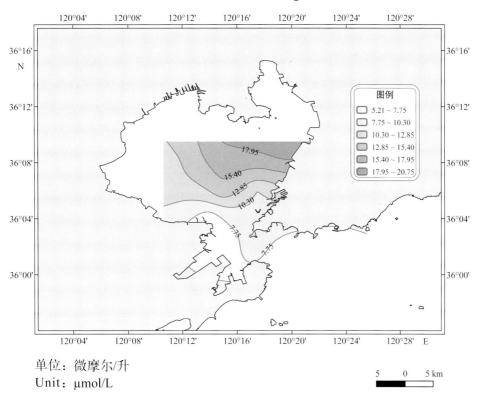

单位：微摩尔/升
Unit：μmol/L

2010年11月硅酸盐分布
Distribution of silicate in November 2010

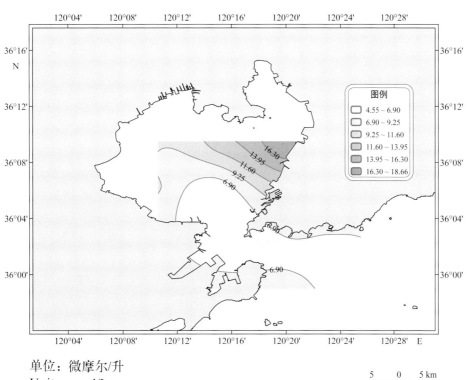

单位：微摩尔/升
Unit：μmol/L

2011年2月硅酸盐分布
Distribution of silicate in February 2011

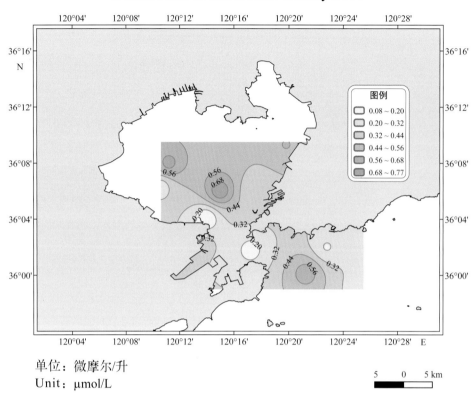

单位：微摩尔/升
Unit：µmol/L

2011年5月硅酸盐分布
Distribution of silicate in May 2011

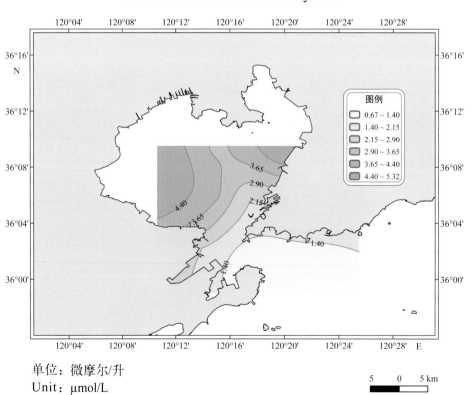

单位：微摩尔/升
Unit：µmol/L

2011年8月硅酸盐分布
Distribution of silicate in August 2011

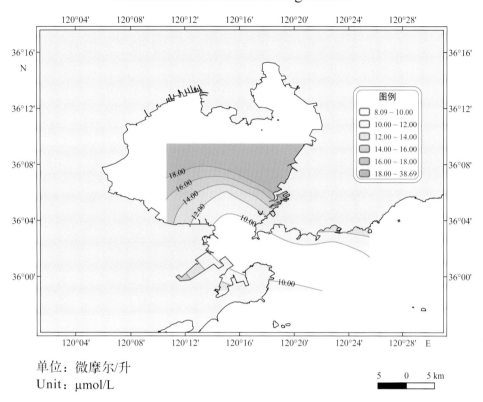

单位：微摩尔/升
Unit：μmol/L

2011年11月硅酸盐分布
Distribution of silicate in November 2011

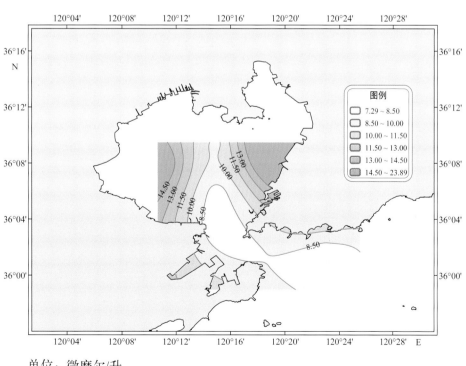

单位：微摩尔/升
Unit：μmol/L

2012年2月硅酸盐分布
Distribution of silicate in February 2012

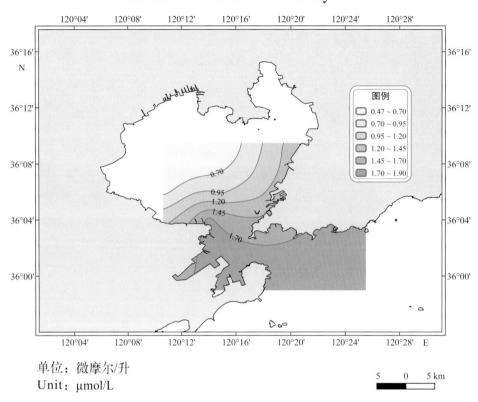

单位：微摩尔/升
Unit：μmol/L

2012年5月硅酸盐分布
Distribution of silicate in May 2012

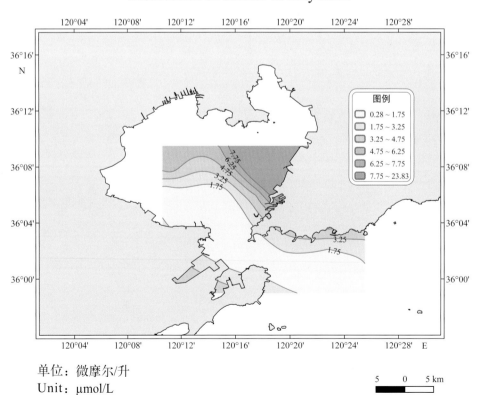

单位：微摩尔/升
Unit：μmol/L

2012年8月硅酸盐分布
Distribution of silicate in August 2012

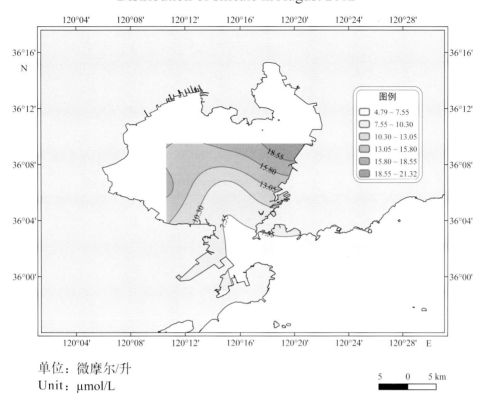

单位：微摩尔/升
Unit：μmol/L

2012年11月硅酸盐分布
Distribution of silicate in November 2012

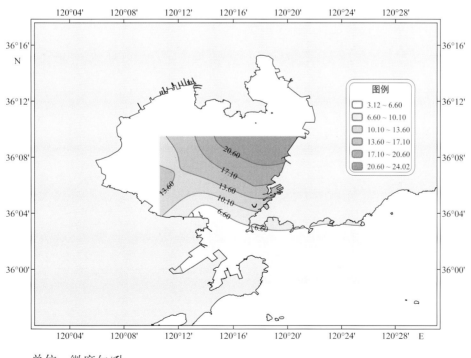

单位：微摩尔/升
Unit：μmol/L

2013 年 2 月硅酸盐分布
Distribution of silicate in February 2013

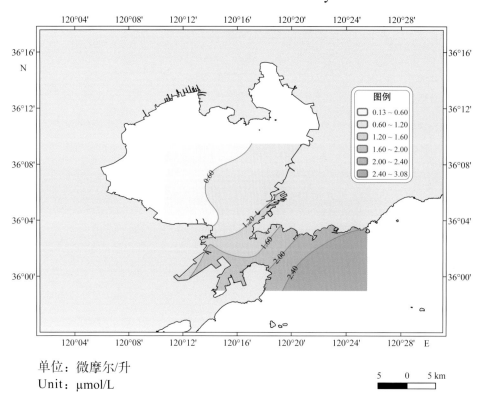

单位：微摩尔/升
Unit：μmol/L

2013 年 5 月硅酸盐分布
Distribution of silicate in May 2013

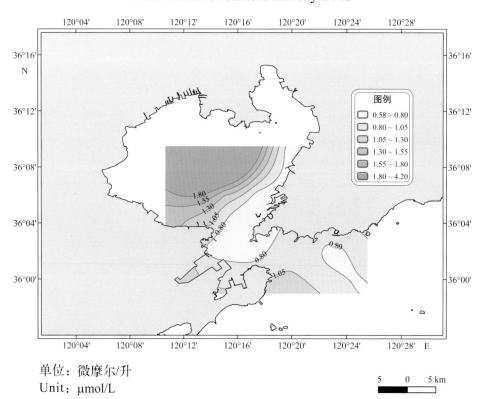

单位：微摩尔/升
Unit：μmol/L

2013年8月硅酸盐分布
Distribution of silicate in August 2013

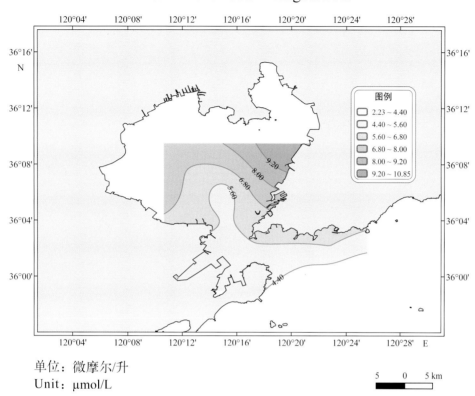

单位：微摩尔/升
Unit：μmol/L

2013年11月硅酸盐分布
Distribution of silicate in November 2013

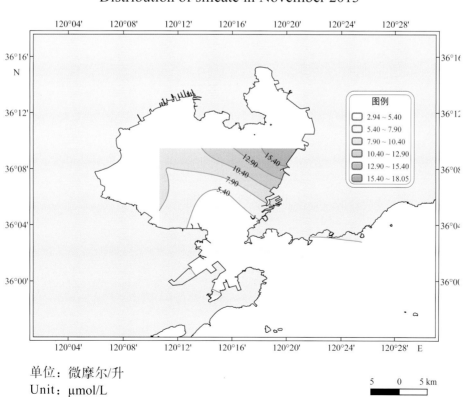

单位：微摩尔/升
Unit：μmol/L

2014年2月硅酸盐分布
Distribution of silicate in February 2014

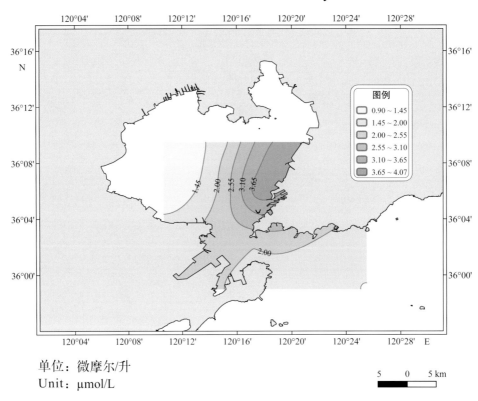

单位：微摩尔/升
Unit：μmol/L

2014年5月硅酸盐分布
Distribution of silicate in May 2014

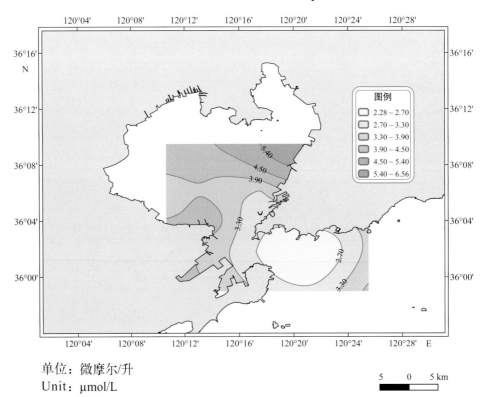

单位：微摩尔/升
Unit：μmol/L

2014年8月硅酸盐分布
Distribution of silicate in August 2014

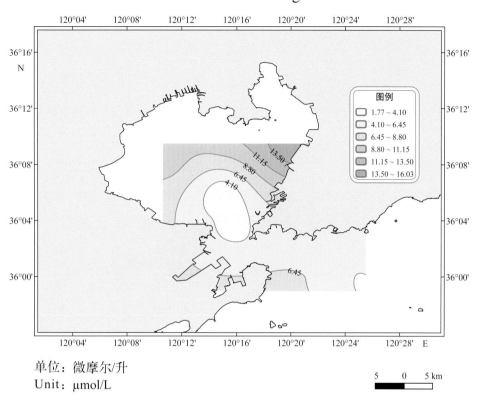

单位：微摩尔/升
Unit：µmol/L

2014年11月硅酸盐分布
Distribution of silicate in November 2014

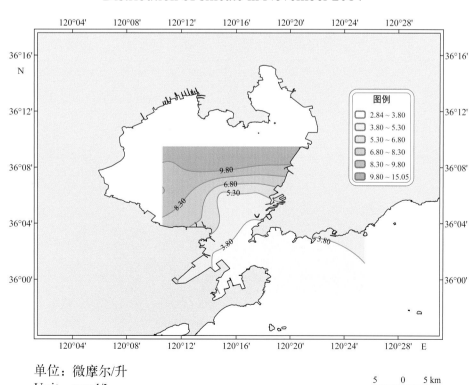

单位：微摩尔/升
Unit：µmol/L

2015年2月硅酸盐分布
Distribution of silicate in February 2015

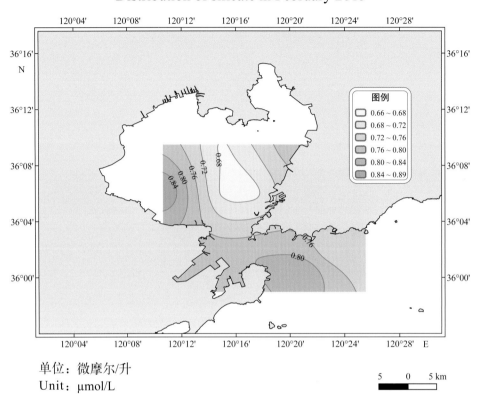

单位：微摩尔/升
Unit：μmol/L

2015年5月硅酸盐分布
Distribution of silicate in May 2015

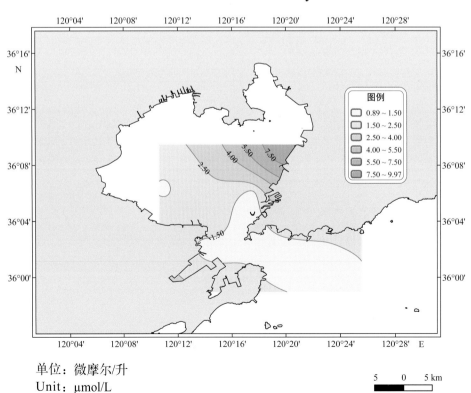

单位：微摩尔/升
Unit：μmol/L

2015年8月硅酸盐分布
Distribution of silicate in August 2015

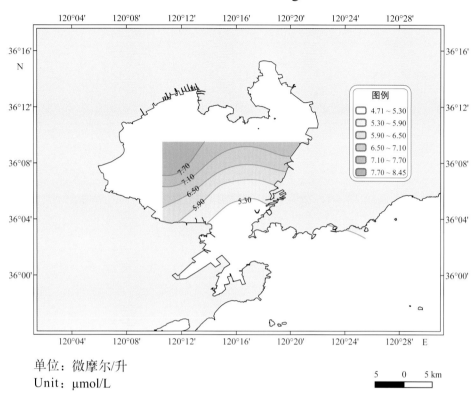

单位：微摩尔/升
Unit：μmol/L

2015年11月硅酸盐分布
Distribution of silicate in November 2015

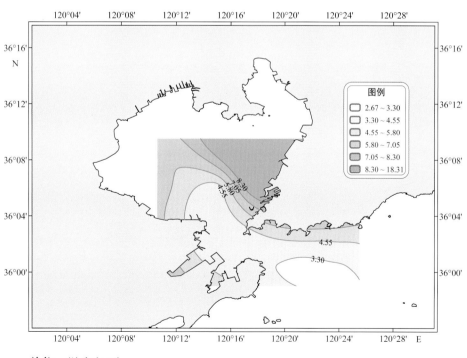

单位：微摩尔/升
Unit：μmol/L

4. 磷酸盐 Phosphate

2010年2月磷酸盐分布
Distribution of phosphate in February 2010

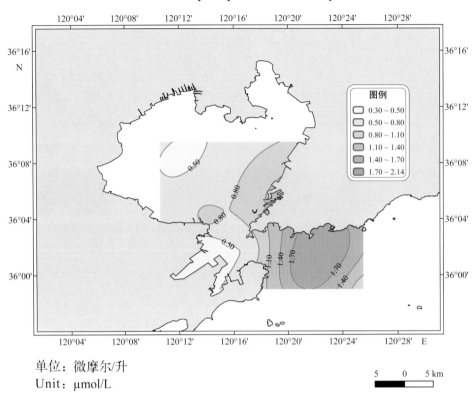

单位：微摩尔/升
Unit：μmol/L

2010年5月磷酸盐分布
Distribution of phosphate in May 2010

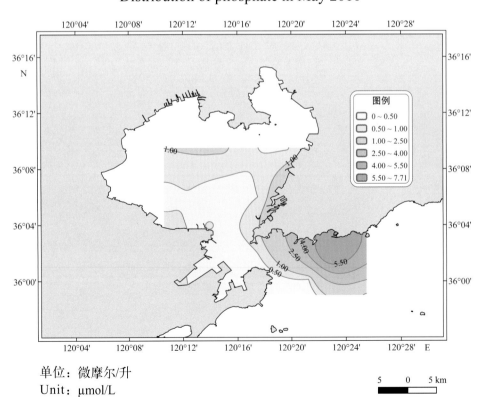

单位：微摩尔/升
Unit：μmol/L

2010年8月磷酸盐分布
Distribution of phosphate in August 2010

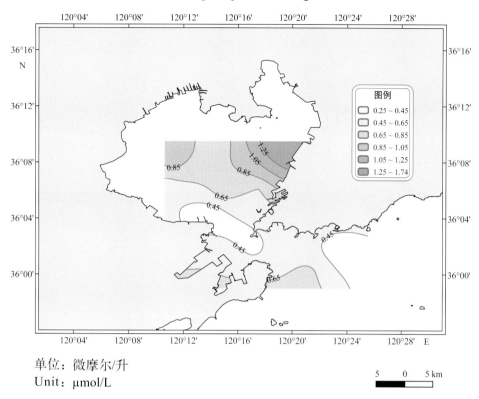

单位：微摩尔/升
Unit：μmol/L

2010年11月磷酸盐分布
Distribution of phosphate in November 2010

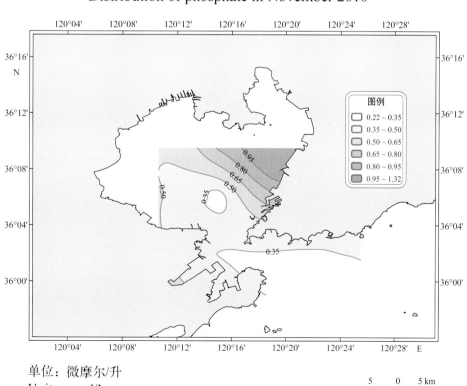

单位：微摩尔/升
Unit：μmol/L

2011年2月磷酸盐分布
Distribution of phosphate in February 2011

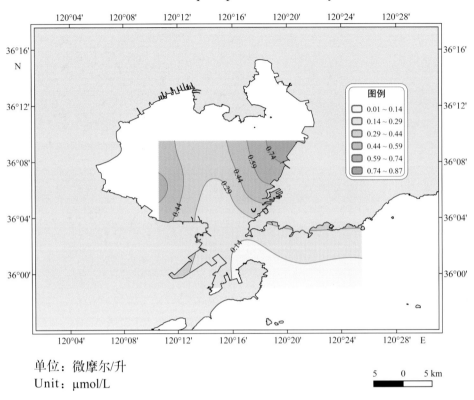

单位:微摩尔/升
Unit: μmol/L

2011年5月磷酸盐分布
Distribution of phosphate in May 2011

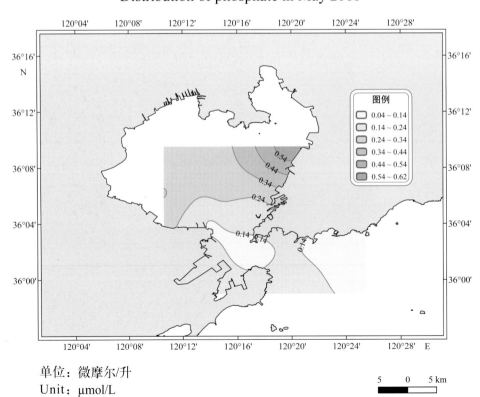

单位:微摩尔/升
Unit: μmol/L

2011年8月磷酸盐分布
Distribution of phosphate in August 2011

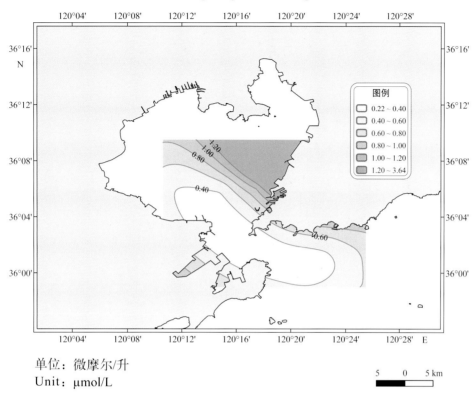

单位：微摩尔/升
Unit：μmol/L

2011年11月磷酸盐分布
Distribution of phosphate in November 2011

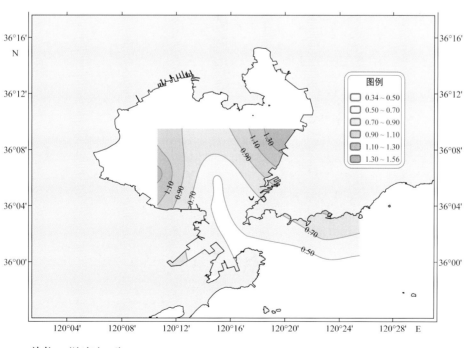

单位：微摩尔/升
Unit：μmol/L

2012年2月磷酸盐分布
Distribution of phosphate in February 2012

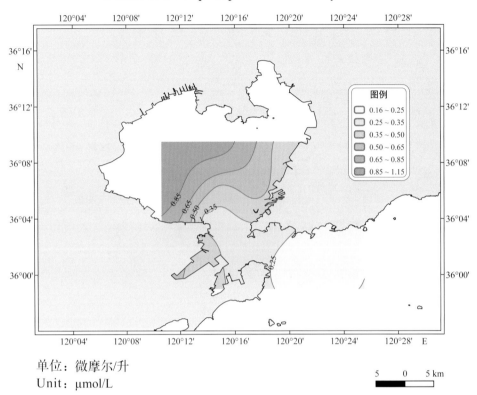

单位：微摩尔/升
Unit：μmol/L

2012年5月磷酸盐分布
Distribution of phosphate in May 2012

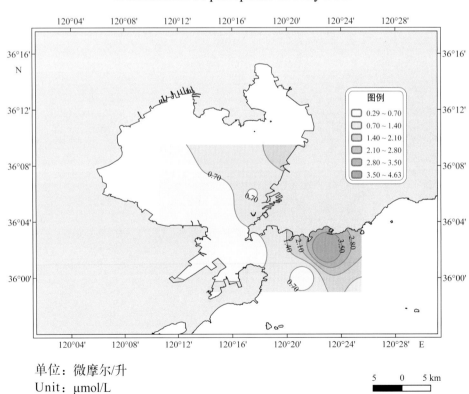

单位：微摩尔/升
Unit：μmol/L

2012年8月磷酸盐分布
Distribution of phosphate in August 2012

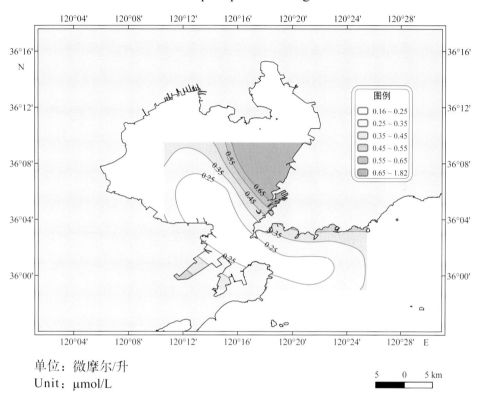

单位：微摩尔/升
Unit：μmol/L

2012年11月磷酸盐分布
Distribution of phosphate in November 2012

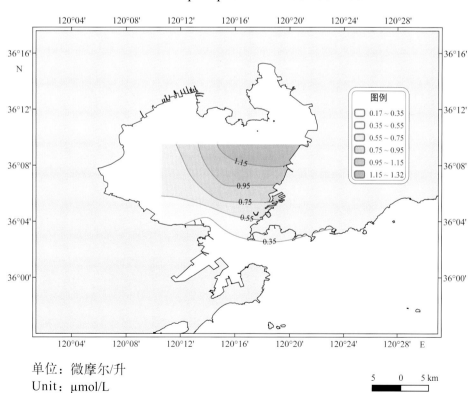

单位：微摩尔/升
Unit：μmol/L

2013年2月磷酸盐分布
Distribution of phosphate in February 2013

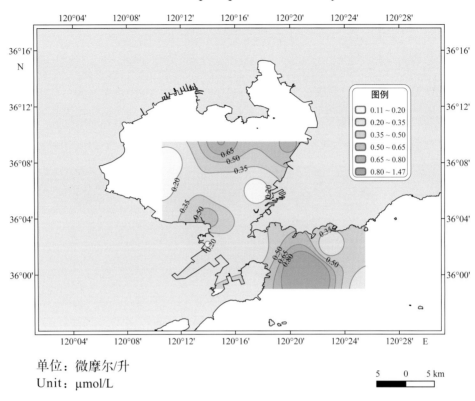

单位：微摩尔/升
Unit：μmol/L

2013年5月磷酸盐分布
Distribution of phosphate in May 2013

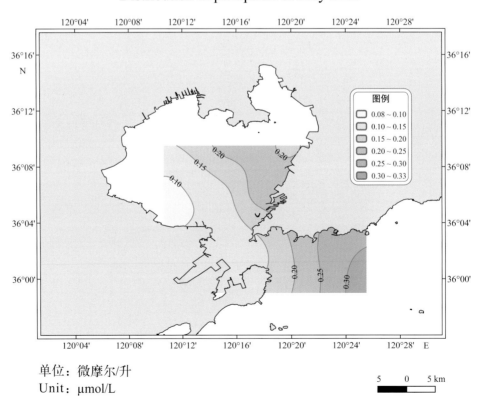

单位：微摩尔/升
Unit：μmol/L

2013年8月磷酸盐分布
Distribution of phosphate in August 2013

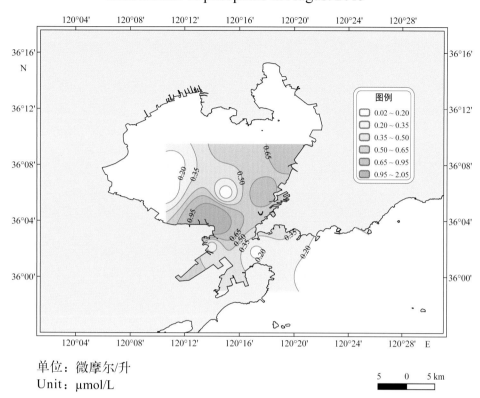

单位：微摩尔/升
Unit：μmol/L

2013年11月磷酸盐分布
Distribution of phosphate in November 2013

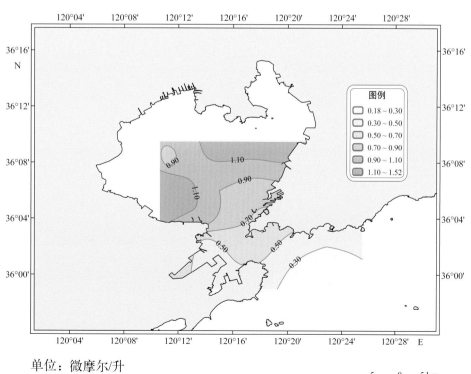

单位：微摩尔/升
Unit：μmol/L

2014年2月磷酸盐分布
Distribution of phosphate in February 2014

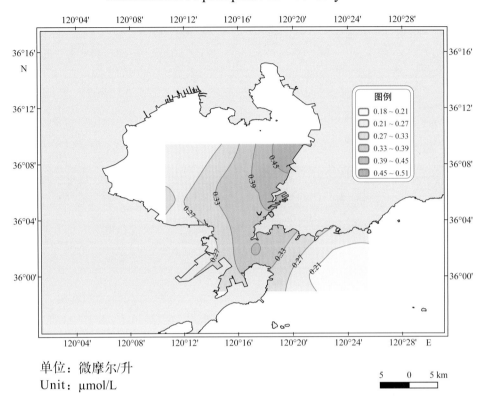

单位：微摩尔/升
Unit：μmol/L

2014年5月磷酸盐分布
Distribution of phosphate in May 2014

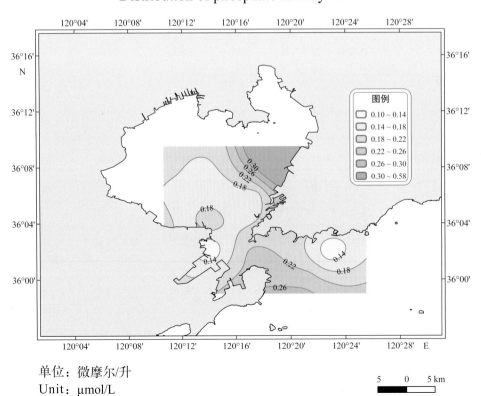

单位：微摩尔/升
Unit：μmol/L

2014年8月磷酸盐分布
Distribution of phosphate in August 2014

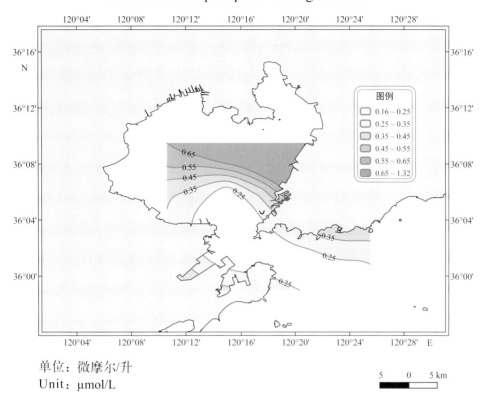

单位：微摩尔/升
Unit：μmol/L

2014年11月磷酸盐分布
Distribution of phosphate in November 2014

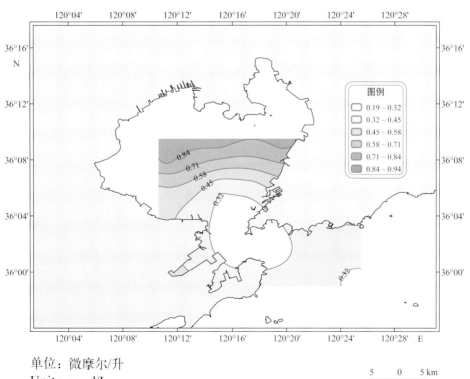

单位：微摩尔/升
Unit：μmol/L

2015年2月磷酸盐分布
Distribution of phosphate in February 2015

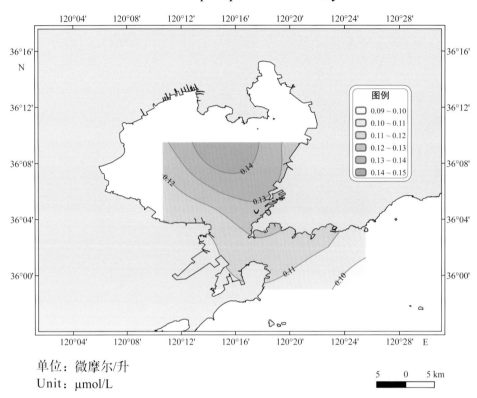

单位：微摩尔/升
Unit：μmol/L

2015年5月磷酸盐分布
Distribution of phosphate in May 2015

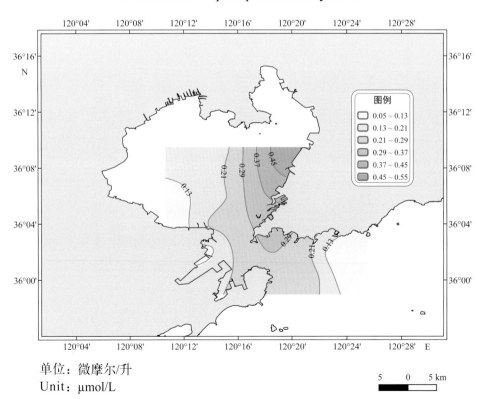

单位：微摩尔/升
Unit：μmol/L

2015年8月磷酸盐分布
Distribution of phosphate in August 2015

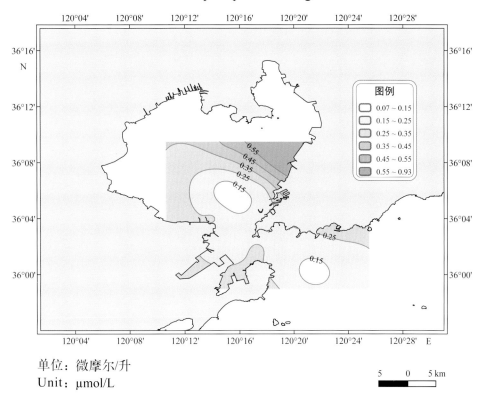

单位：微摩尔/升
Unit：μmol/L

2015年11月磷酸盐分布
Distribution of phosphate in November 2015

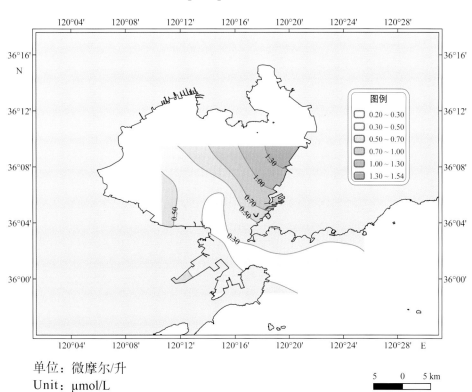

单位：微摩尔/升
Unit：μmol/L

5. 亚硝酸盐 Nitrite

2010年2月亚硝酸盐分布
Distribution of nitrite in February 2010

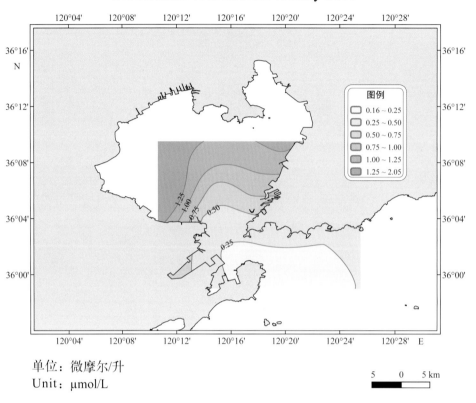

单位：微摩尔/升
Unit：μmol/L

2010年5月亚硝酸盐分布
Distribution of nitrite in May 2010

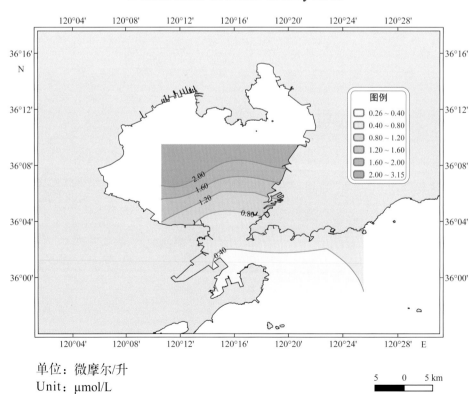

单位：微摩尔/升
Unit：μmol/L

2010年8月亚硝酸盐分布
Distribution of nitrite in August 2010

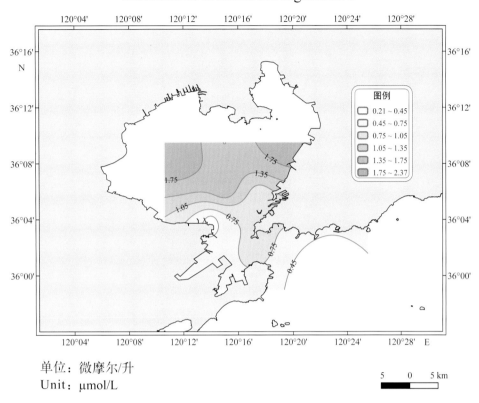

单位：微摩尔/升
Unit：μmol/L

2010年11月亚硝酸盐分布
Distribution of nitrite in November 2010

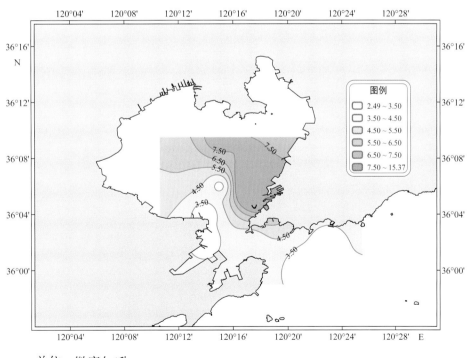

单位：微摩尔/升
Unit：μmol/L

2011年2月亚硝酸盐分布
Distribution of nitrite in February 2011

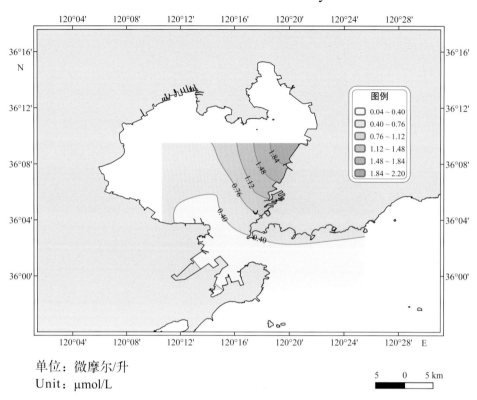

单位：微摩尔/升
Unit：µmol/L

2011年5月亚硝酸盐分布
Distribution of nitrite in May 2011

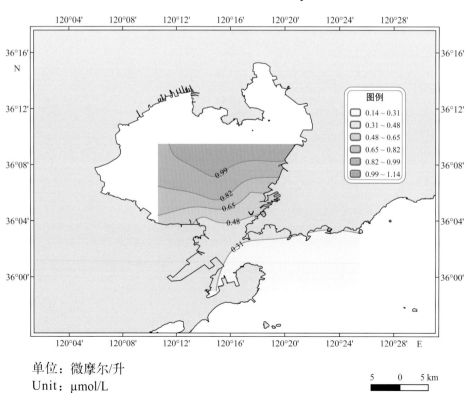

单位：微摩尔/升
Unit：µmol/L

2011年8月亚硝酸盐分布
Distribution of nitrite in August 2011

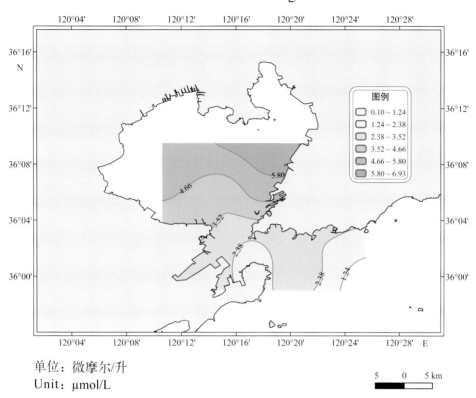

单位：微摩尔/升
Unit：μmol/L

2011年11月亚硝酸盐分布
Distribution of nitrite in November 2011

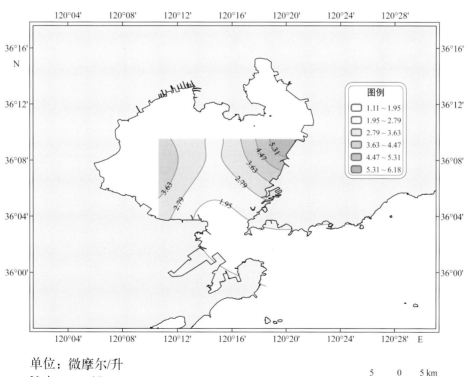

单位：微摩尔/升
Unit：μmol/L

2012年2月亚硝酸盐分布
Distribution of nitrite in February 2012

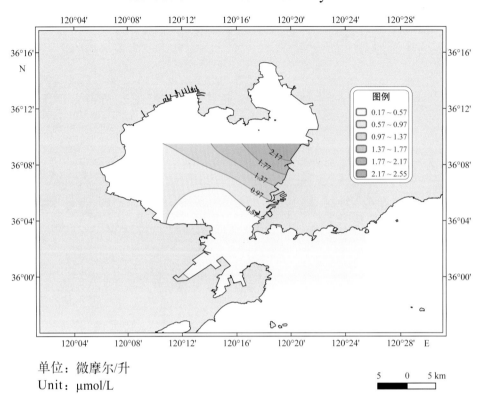

单位：微摩尔/升
Unit：µmol/L

2012年5月亚硝酸盐分布
Distribution of nitrite in May 2012

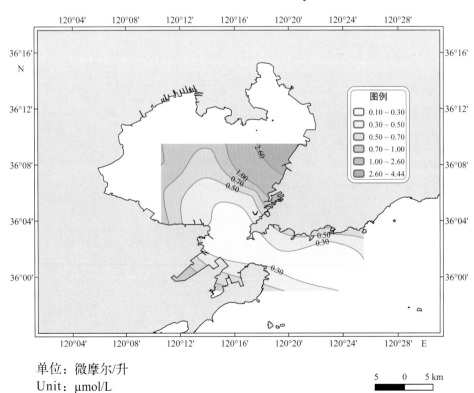

单位：微摩尔/升
Unit：µmol/L

2012年8月亚硝酸盐分布
Distribution of nitrite in August 2012

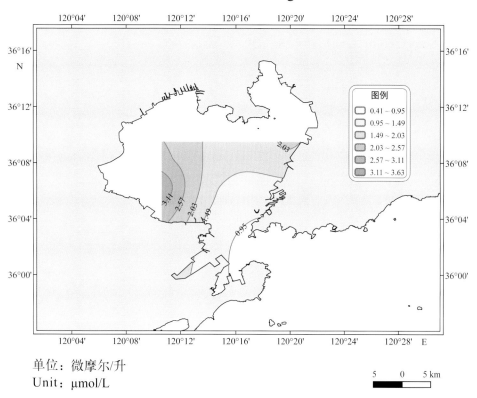

单位：微摩尔/升
Unit：μmol/L

2012年11月亚硝酸盐分布
Distribution of nitrite in November 2012

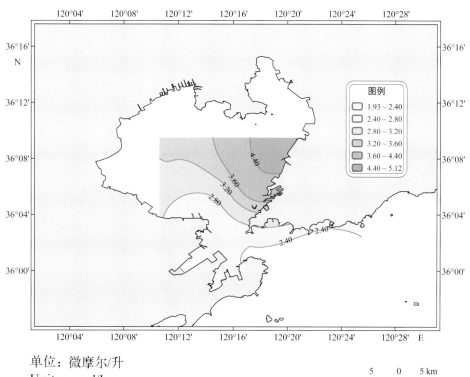

单位：微摩尔/升
Unit：μmol/L

2013年2月亚硝酸盐分布
Distribution of nitrite in February 2013

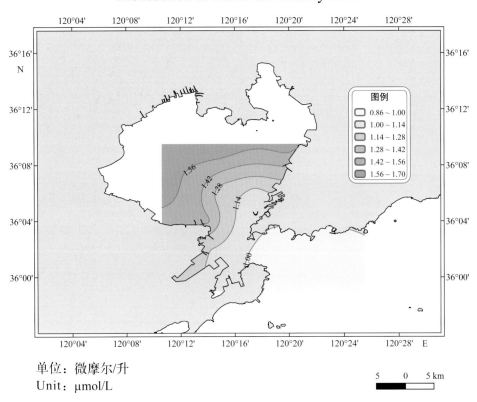

单位：微摩尔/升
Unit：μmol/L

2013年5月亚硝酸盐分布
Distribution of nitrite in May 2013

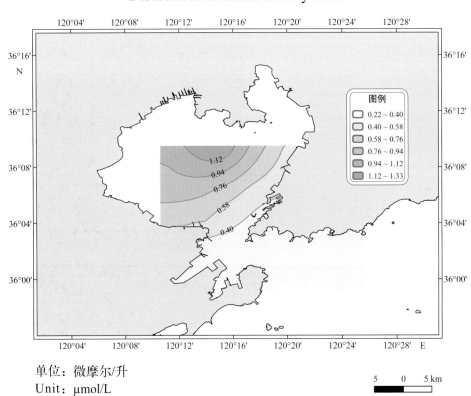

单位：微摩尔/升
Unit：μmol/L

2013年8月亚硝酸盐分布
Distribution of nitrite in August 2013

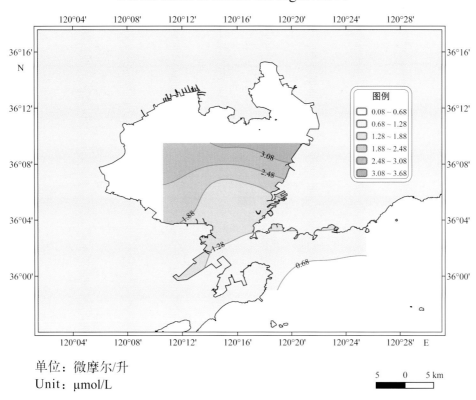

单位：微摩尔/升
Unit：μmol/L

2013年11月亚硝酸盐分布
Distribution of nitrite in November 2013

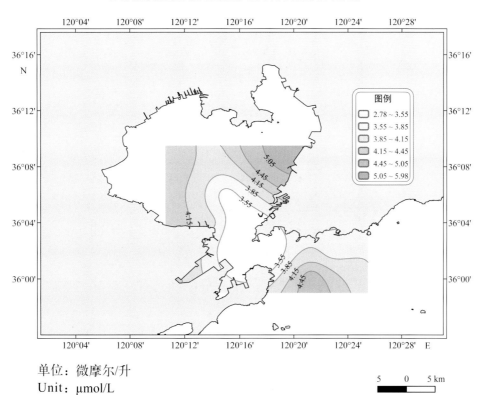

单位：微摩尔/升
Unit：μmol/L

2014年2月亚硝酸盐分布
Distribution of nitrite in February 2014

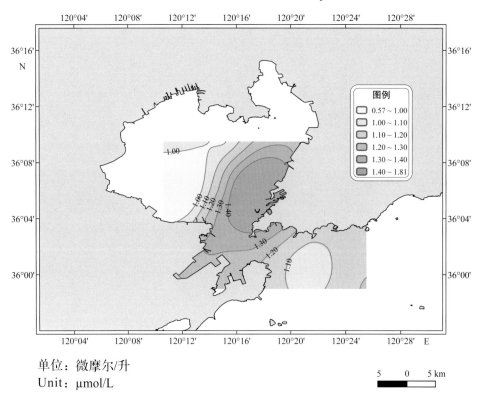

单位：微摩尔/升
Unit：μmol/L

2014年5月亚硝酸盐分布
Distribution of nitrite in May 2014

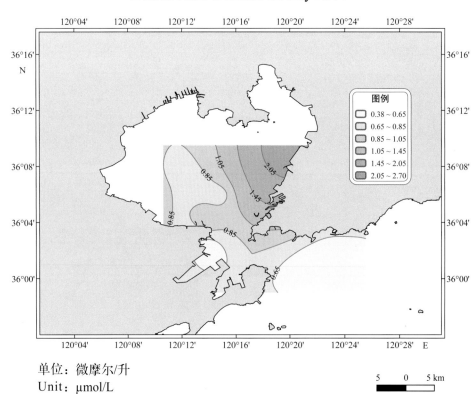

单位：微摩尔/升
Unit：μmol/L

2014年8月亚硝酸盐分布
Distribution of nitrite in August 2014

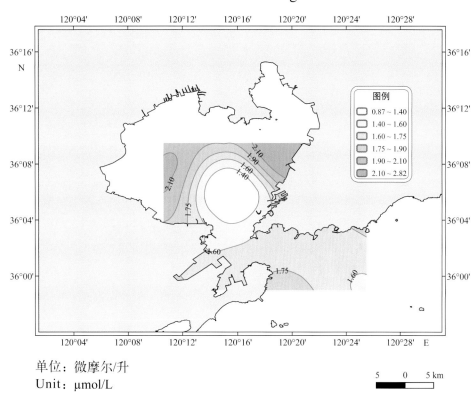

单位：微摩尔/升
Unit：µmol/L

2014年11月亚硝酸盐分布
Distribution of nitrite in November 2014

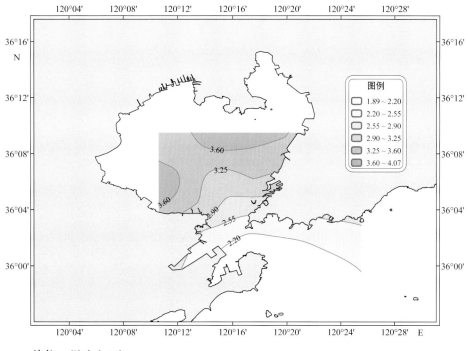

单位：微摩尔/升
Unit：µmol/L

2015年2月亚硝酸盐分布
Distribution of nitrite in February 2015

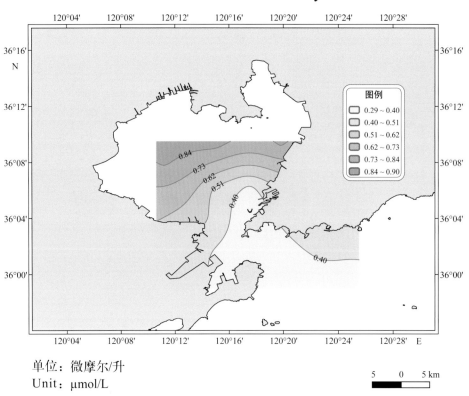

单位：微摩尔/升
Unit：μmol/L

2015年5月亚硝酸盐分布
Distribution of nitrite in May 2015

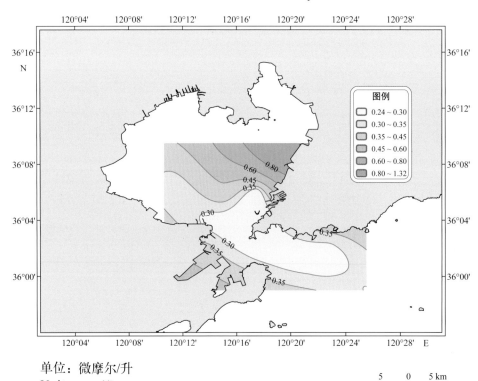

单位：微摩尔/升
Unit：μmol/L

2015年8月亚硝酸盐分布
Distribution of nitrite in August 2015

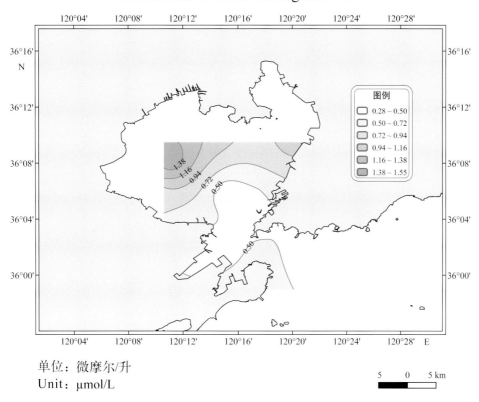

单位：微摩尔/升
Unit：μmol/L

2015年11月亚硝酸盐分布
Distribution of nitrite in November 2015

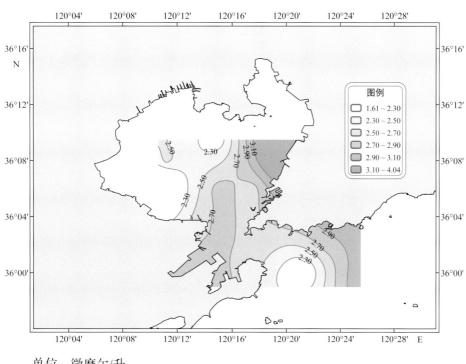

单位：微摩尔/升
Unit：μmol/L

6. 硝酸盐 Nitrate

2010年2月硝酸盐分布
Distribution of nitrate in February 2010

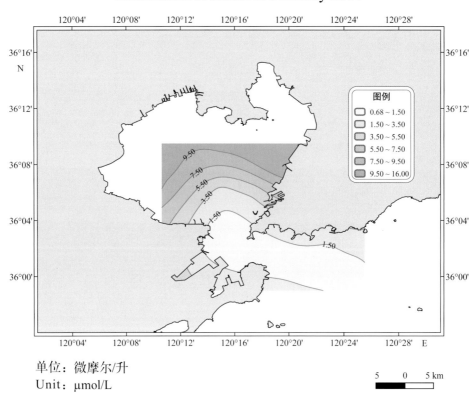

单位：微摩尔/升
Unit：µmol/L

2010年5月硝酸盐分布
Distribution of nitrate in May 2010

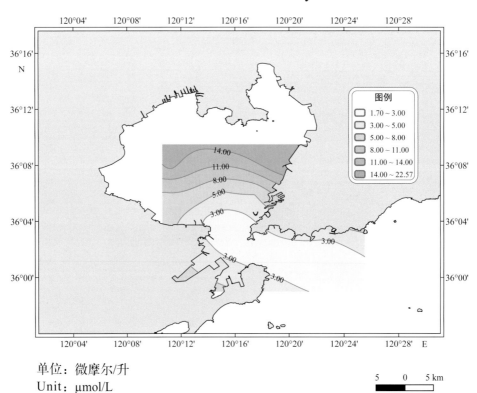

单位：微摩尔/升
Unit：µmol/L

2010年8月硝酸盐分布
Distribution of nitrate in August 2010

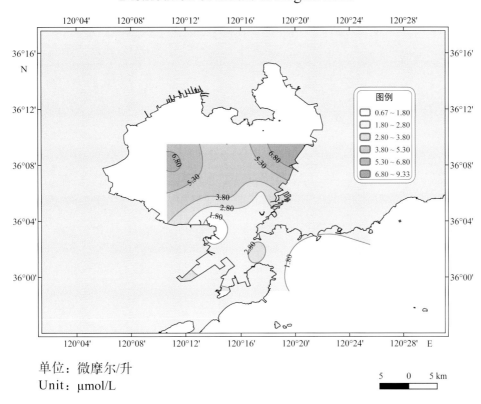

单位：微摩尔/升
Unit：μmol/L

2010年11月硝酸盐分布
Distribution of nitrate in November 2010

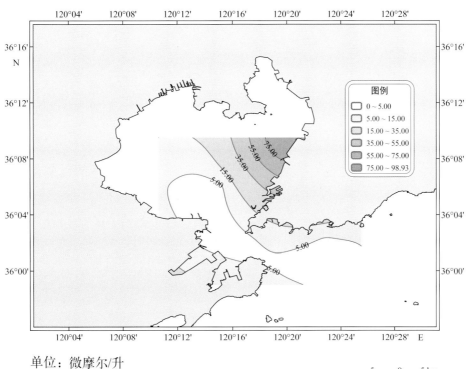

单位：微摩尔/升
Unit：μmol/L

2011年2月硝酸盐分布
Distribution of nitrate in February 2011

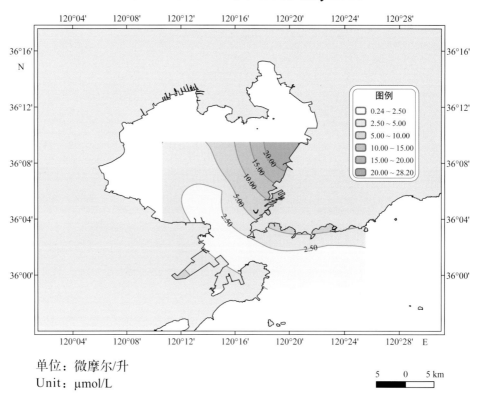

单位：微摩尔/升
Unit：μmol/L

2011年5月硝酸盐分布
Distribution of nitrate in May 2011

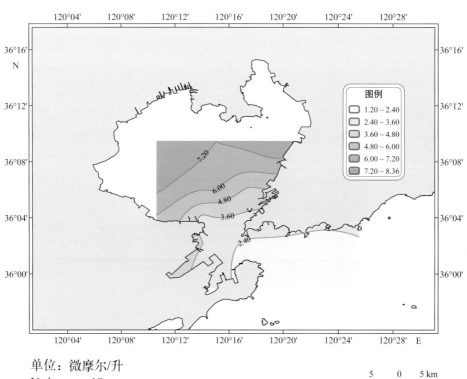

单位：微摩尔/升
Unit：μmol/L

2011年8月硝酸盐分布
Distribution of nitrate in August 2011

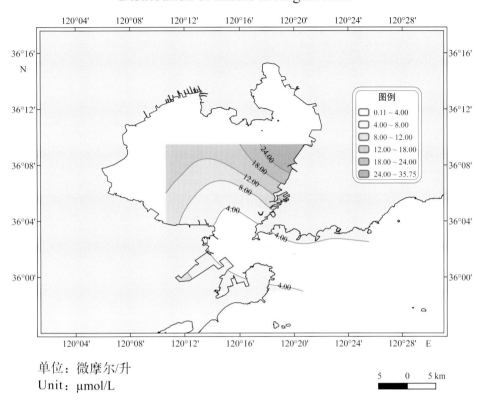

单位：微摩尔/升
Unit：μmol/L

2011年11月硝酸盐分布
Distribution of nitrate in November 2011

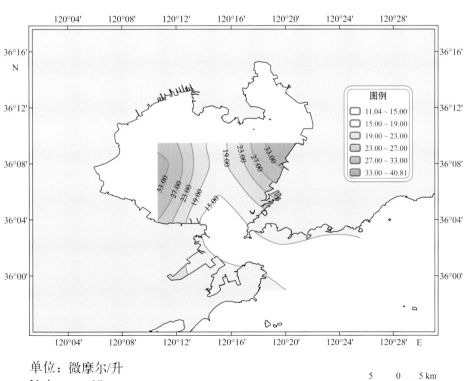

单位：微摩尔/升
Unit：μmol/L

2012年2月硝酸盐分布
Distribution of nitrate in February 2012

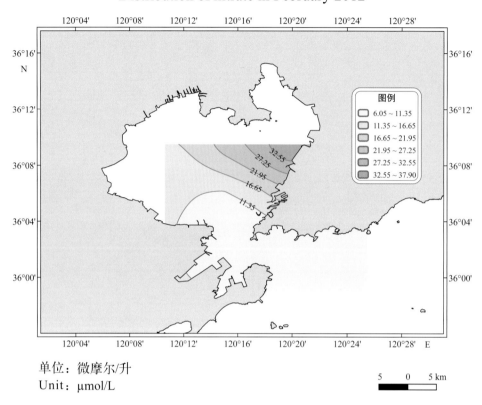

单位：微摩尔/升
Unit：μmol/L

2012年5月硝酸盐分布
Distribution of nitrate in May 2012

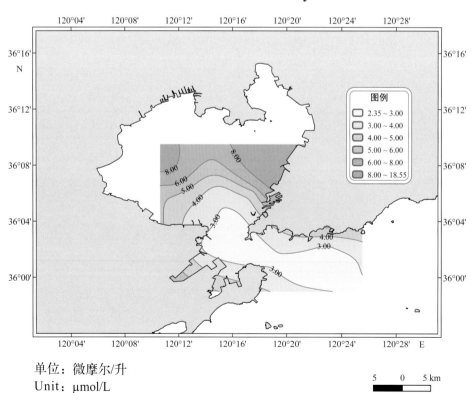

单位：微摩尔/升
Unit：μmol/L

2012年8月硝酸盐分布
Distribution of nitrate in August 2012

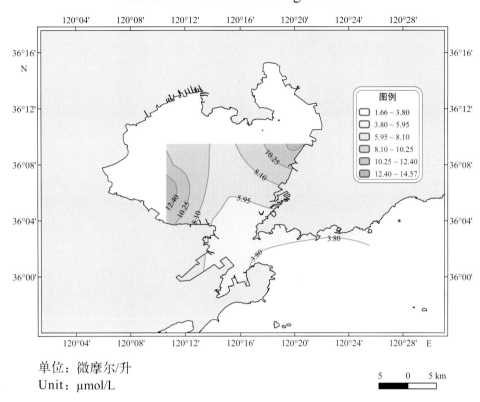

单位：微摩尔/升
Unit：μmol/L

2012年11月硝酸盐分布
Distribution of nitrate in November 2012

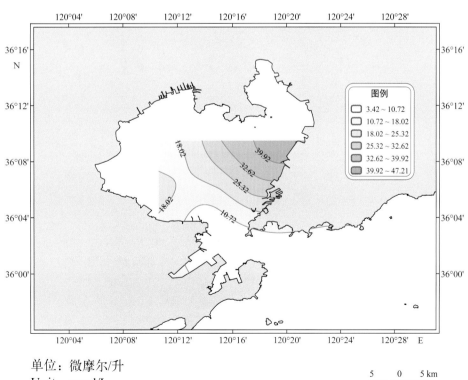

单位：微摩尔/升
Unit：μmol/L

2013年2月硝酸盐分布
Distribution of nitrate in February 2013

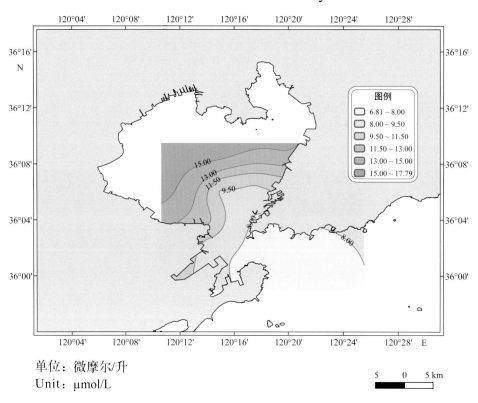

单位：微摩尔/升
Unit：μmol/L

2013年5月硝酸盐分布
Distribution of nitrate in May 2013

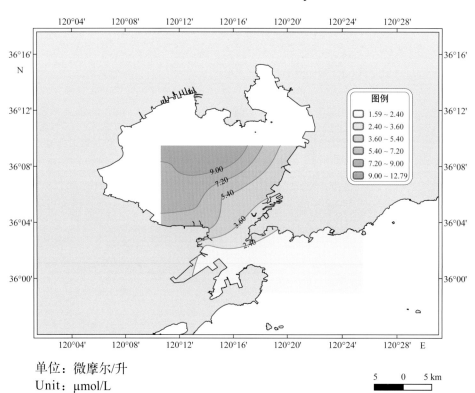

单位：微摩尔/升
Unit：μmol/L

2013年8月硝酸盐分布
Distribution of nitrate in August 2013

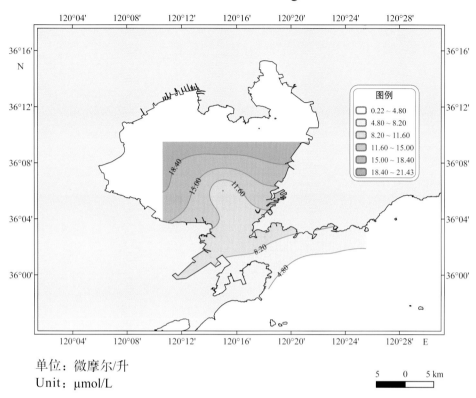

单位：微摩尔/升
Unit：μmol/L

2013年11月硝酸盐分布
Distribution of nitrate in November 2013

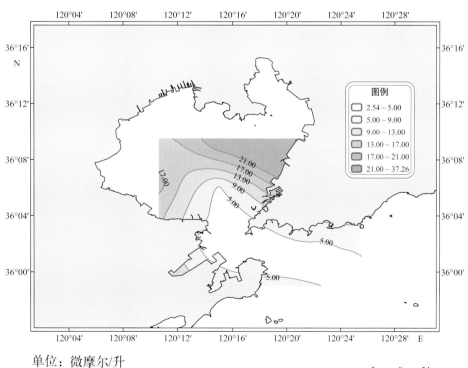

单位：微摩尔/升
Unit：μmol/L

2014年2月硝酸盐分布
Distribution of nitrate in February 2014

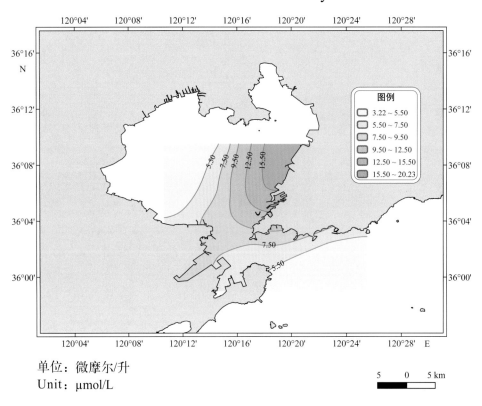

单位：微摩尔/升
Unit：μmol/L

2014年5月硝酸盐分布
Distribution of nitrate in May 2014

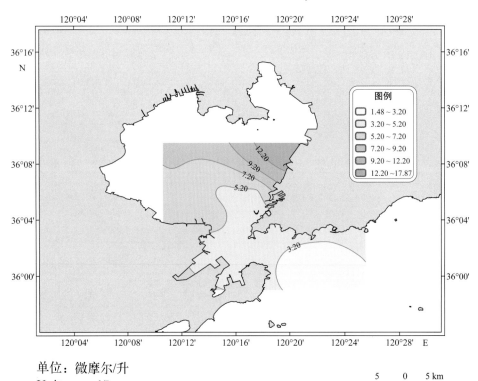

单位：微摩尔/升
Unit：μmol/L

2014年8月硝酸盐分布
Distribution of nitrate in August 2014

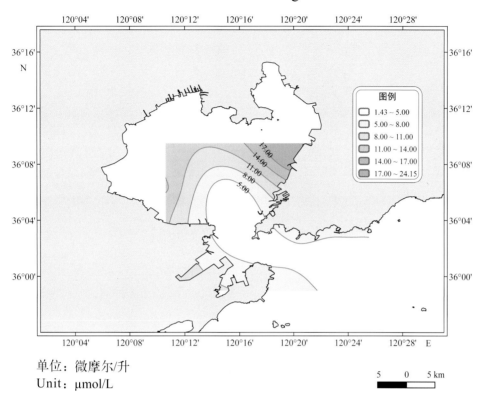

单位:微摩尔/升
Unit: μmol/L

2014年11月硝酸盐分布
Distribution of nitrate in November 2014

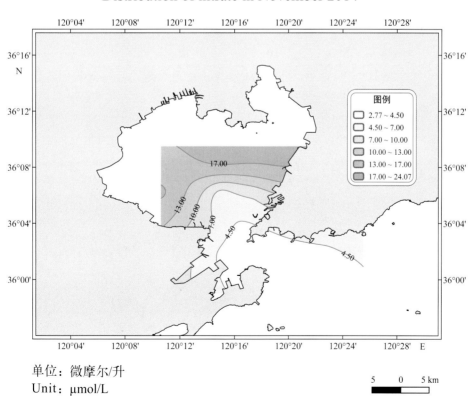

单位:微摩尔/升
Unit: μmol/L

2015年2月硝酸盐分布
Distribution of nitrate in February 2015

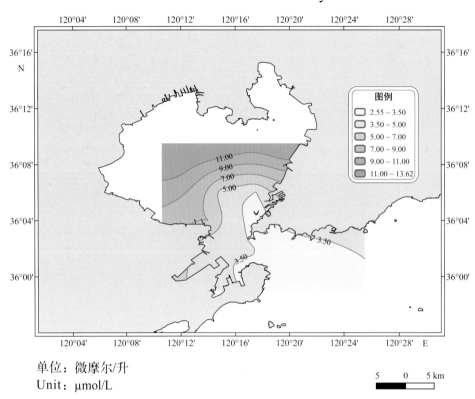

单位：微摩尔/升
Unit：μmol/L

2015年5月硝酸盐分布
Distribution of nitrate in May 2015

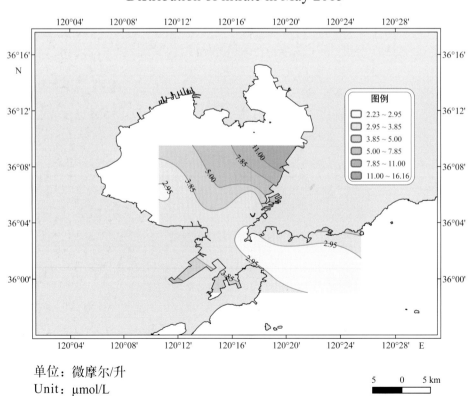

单位：微摩尔/升
Unit：μmol/L

2015年8月硝酸盐分布
Distribution of nitrate in August 2015

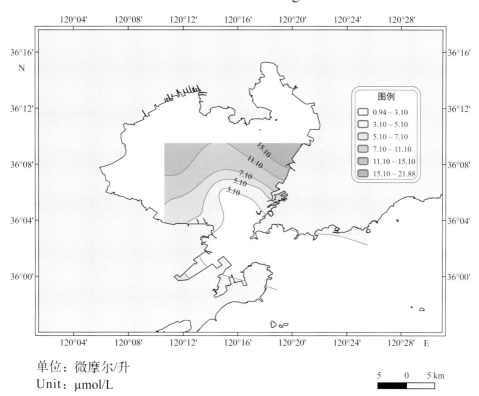

单位：微摩尔/升
Unit：µmol/L

2015年11月硝酸盐分布
Distribution of nitrate in November 2015

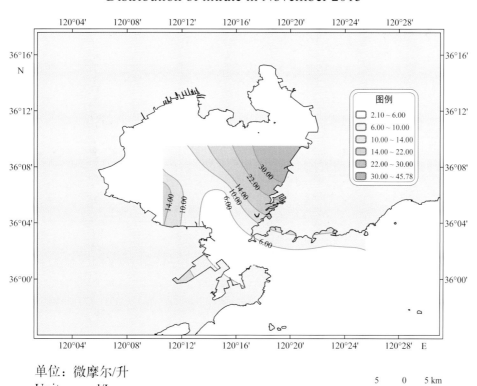

单位：微摩尔/升
Unit：µmol/L

7. 铵盐 Ammonium

2010年2月铵盐分布
Distribution of ammonium in February 2010

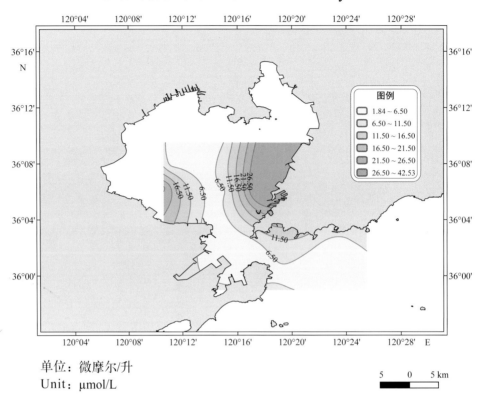

单位：微摩尔/升
Unit：μmol/L

2010年5月铵盐分布
Distribution of ammonium in May 2010

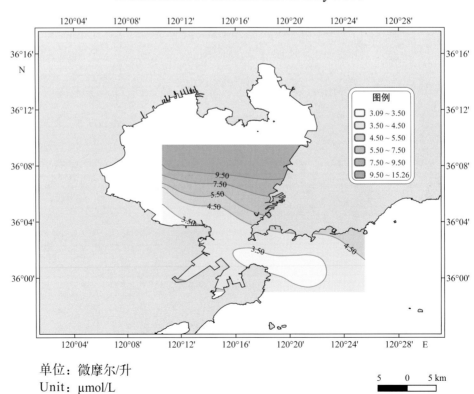

单位：微摩尔/升
Unit：μmol/L

2010年8月铵盐分布
Distribution of ammonium in August 2010

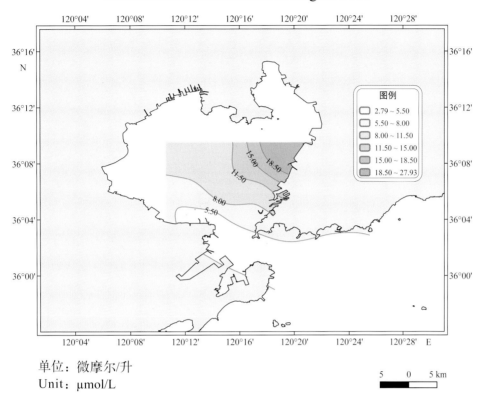

单位：微摩尔/升
Unit：μmol/L

2010年11月铵盐分布
Distribution of ammonium in November 2010

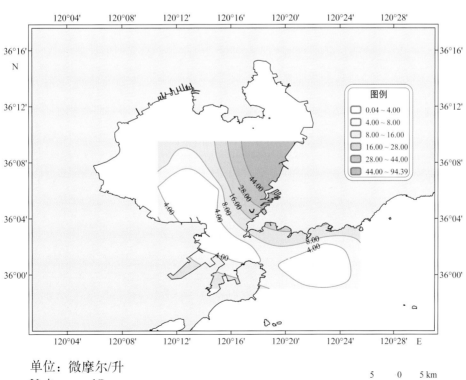

单位：微摩尔/升
Unit：μmol/L

2011年2月铵盐分布
Distribution of ammonium in February 2011

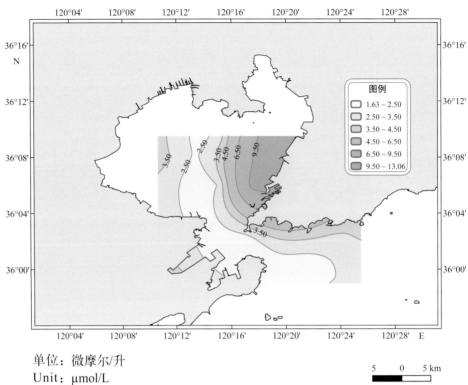

单位：微摩尔/升
Unit：µmol/L

2011年5月铵盐分布
Distribution of ammonium in May 2011

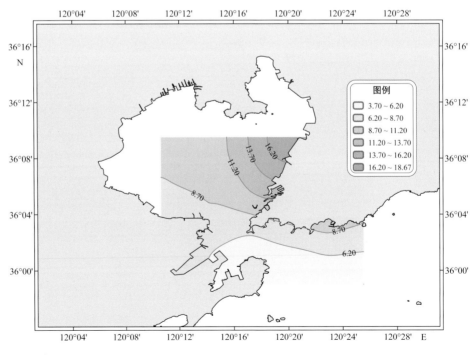

单位：微摩尔/升
Unit：µmol/L

2011年8月铵盐分布
Distribution of ammonium in August 2011

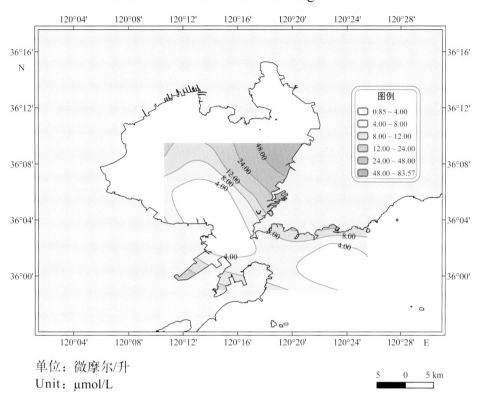

单位：微摩尔/升
Unit：μmol/L

2011年11月铵盐分布
Distribution of ammonium in November 2011

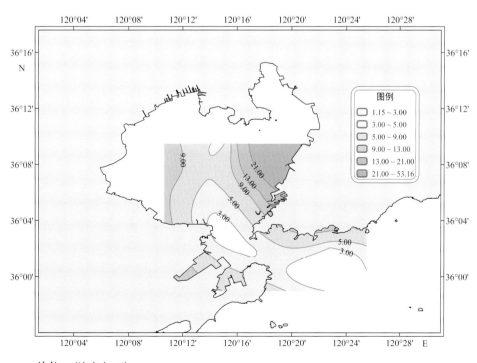

单位：微摩尔/升
Unit：μmol/L

2012年2月铵盐分布
Distribution of ammonium in February 2012

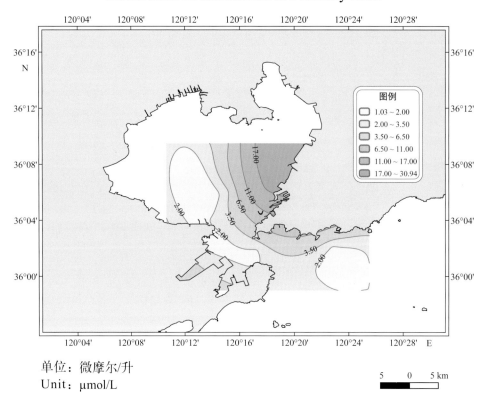

单位：微摩尔/升
Unit：µmol/L

2012年5月铵盐分布
Distribution of ammonium in May 2012

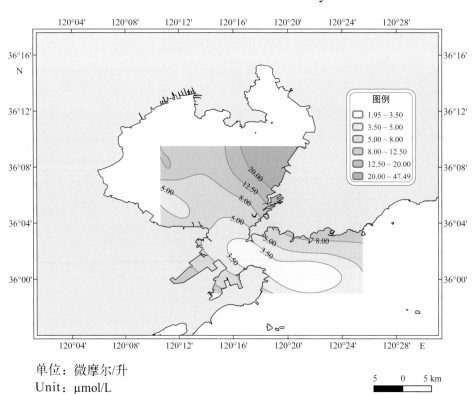

单位：微摩尔/升
Unit：µmol/L

2012年8月铵盐分布
Distribution of ammonium in August 2012

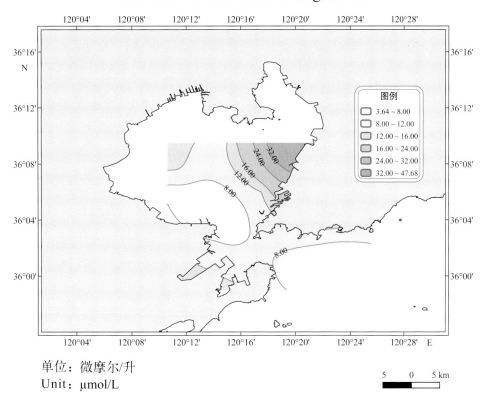

单位：微摩尔/升
Unit：μmol/L

2012年11月铵盐分布
Distribution of ammonium in November 2012

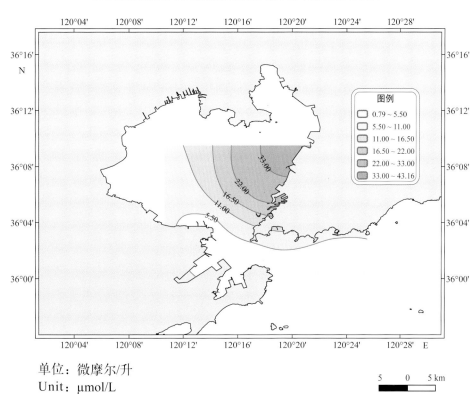

单位：微摩尔/升
Unit：μmol/L

2013年2月铵盐分布
Distribution of ammonium in February 2013

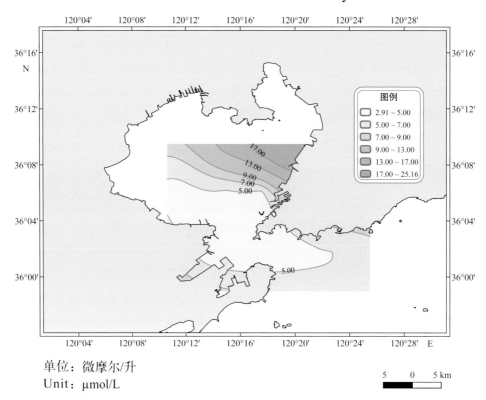

单位：微摩尔/升
Unit：µmol/L

2013年5月铵盐分布
Distribution of ammonium in May 2013

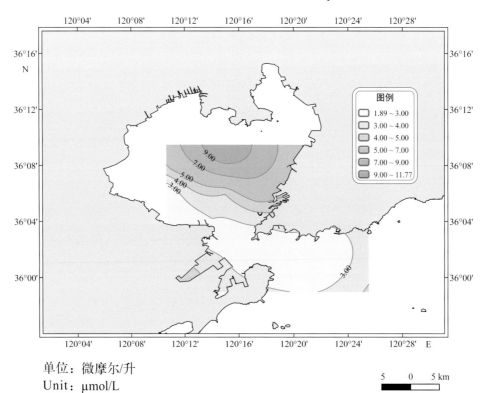

单位：微摩尔/升
Unit：µmol/L

2013年8月铵盐分布
Distribution of ammonium in August 2013

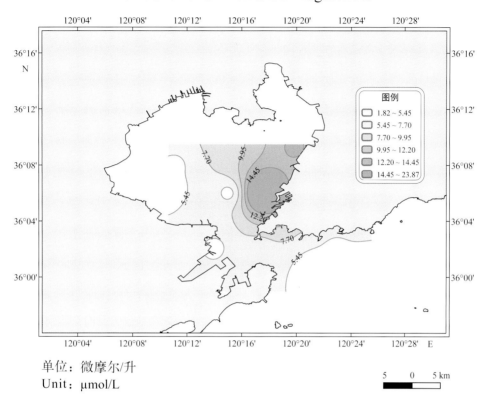

单位：微摩尔/升
Unit：μmol/L

2013年11月铵盐分布
Distribution of ammonium in November 2013

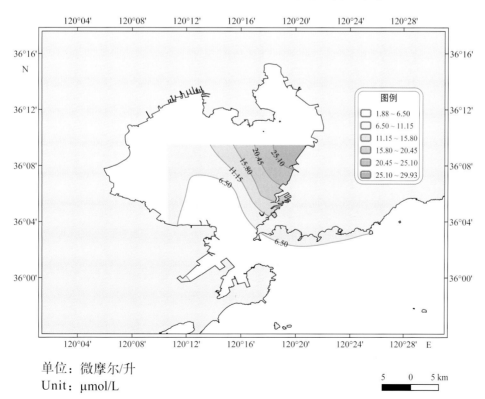

单位：微摩尔/升
Unit：μmol/L

2014 年 2 月铵盐分布
Distribution of ammonium in February 2014

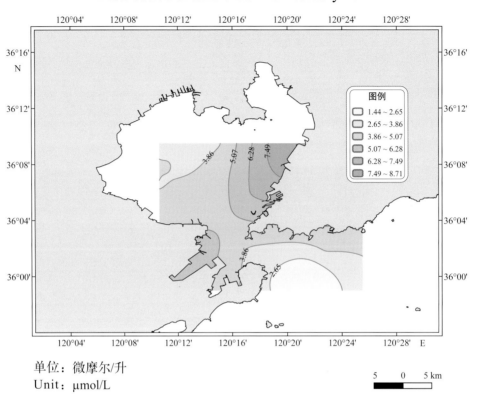

单位：微摩尔/升
Unit：µmol/L

2014 年 5 月铵盐分布
Distribution of ammonium in May 2014

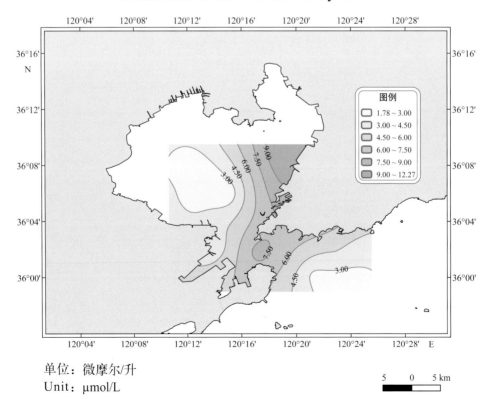

单位：微摩尔/升
Unit：µmol/L

2014年8月铵盐分布
Distribution of ammonium in August 2014

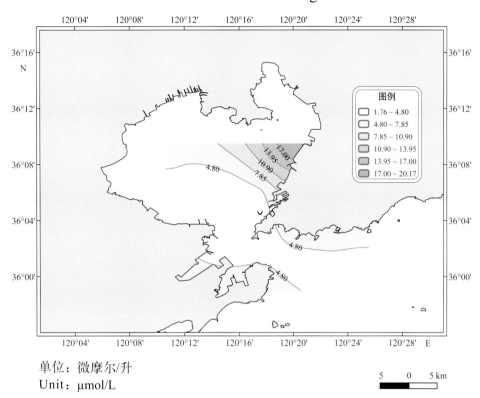

单位：微摩尔/升
Unit：μmol/L

2014年11月铵盐分布
Distribution of ammonium in November 2014

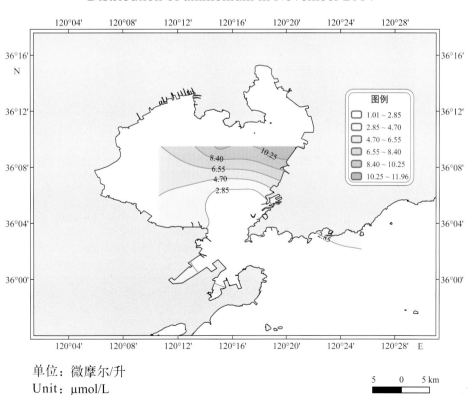

单位：微摩尔/升
Unit：μmol/L

2015年2月铵盐分布
Distribution of ammonium in February 2015

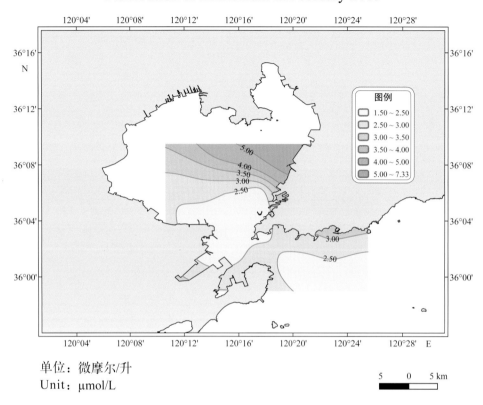

单位：微摩尔/升
Unit：μmol/L

2015年5月铵盐分布
Distribution of ammonium in May 2015

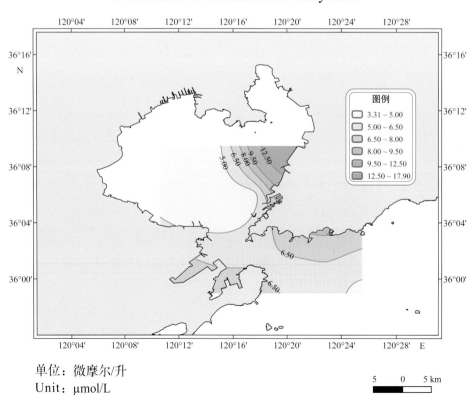

单位：微摩尔/升
Unit：μmol/L

2015年8月铵盐分布
Distribution of ammonium in August 2015

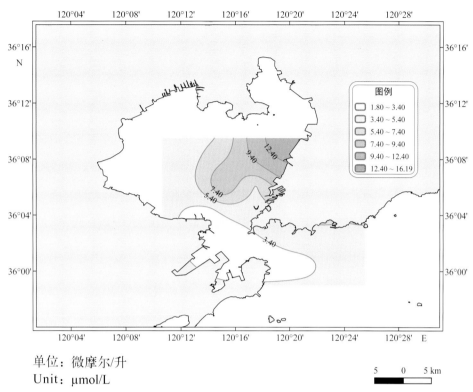

单位：微摩尔/升
Unit：μmol/L

2015年11月铵盐分布
Distribution of ammonium in November 2015

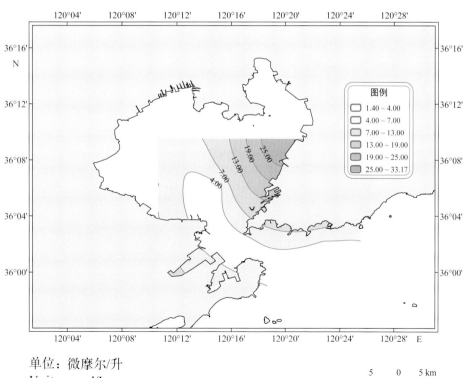

单位：微摩尔/升
Unit：μmol/L

8. 溶解有机碳 Dissolved organic carbon

2010年2月溶解有机碳分布
Distribution of dissolved organic carbon in February 2010

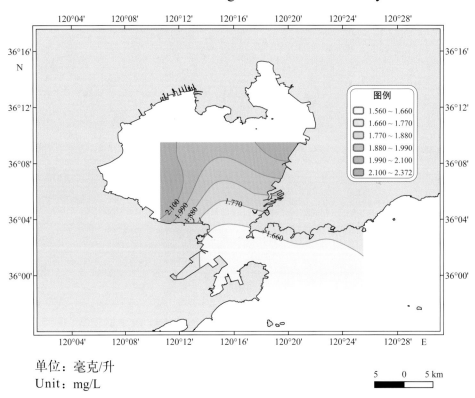

单位：毫克/升
Unit：mg/L

2010年5月溶解有机碳分布
Distribution of dissolved organic carbon in May 2010

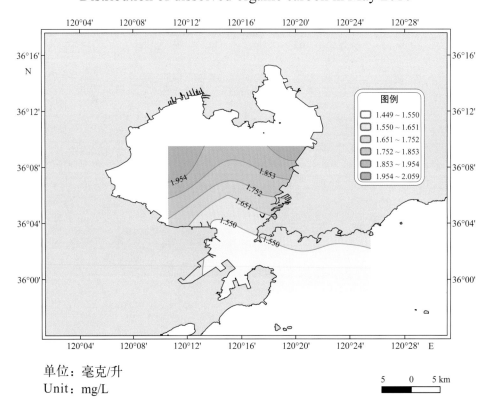

单位：毫克/升
Unit：mg/L

2010年8月溶解有机碳分布
Distribution of dissolved organic carbon in August 2010

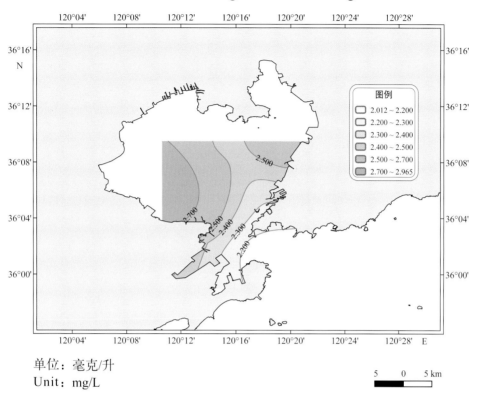

单位：毫克/升
Unit：mg/L

2010年11月溶解有机碳分布
Distribution of dissolved organic carbon in November 2010

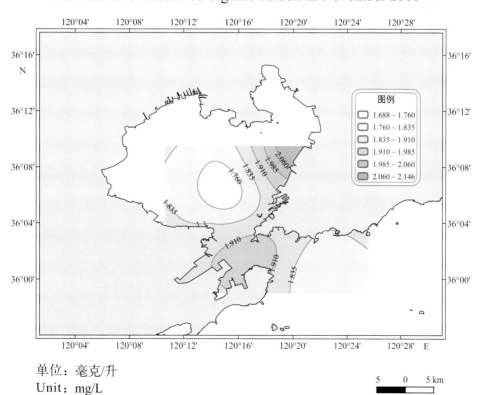

单位：毫克/升
Unit：mg/L

2011年2月溶解有机碳分布
Distribution of dissolved organic carbon in February 2011

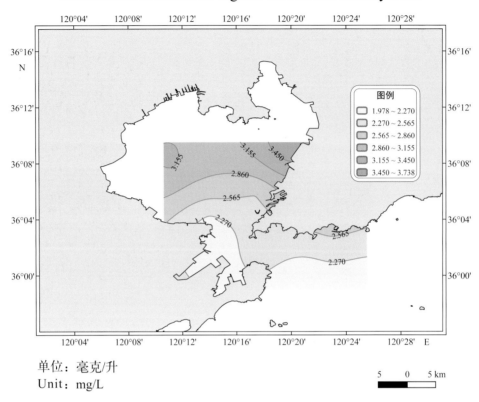

单位：毫克/升
Unit：mg/L

2011年5月溶解有机碳分布
Distribution of dissolved organic carbon in May 2011

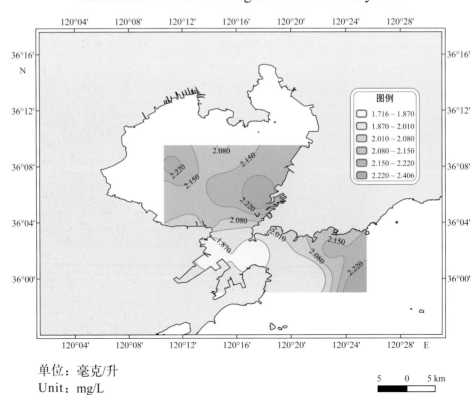

单位：毫克/升
Unit：mg/L

2011年8月溶解有机碳分布
Distribution of dissolved organic carbon in August 2011

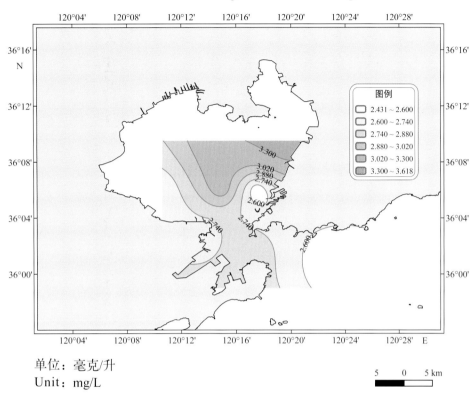

单位：毫克/升
Unit：mg/L

2011年11月溶解有机碳分布
Distribution of dissolved organic carbon in November 2011

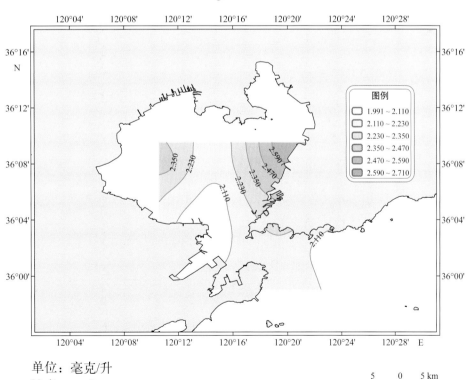

单位：毫克/升
Unit：mg/L

2012年2月溶解有机碳分布
Distribution of dissolved organic carbon in February 2012

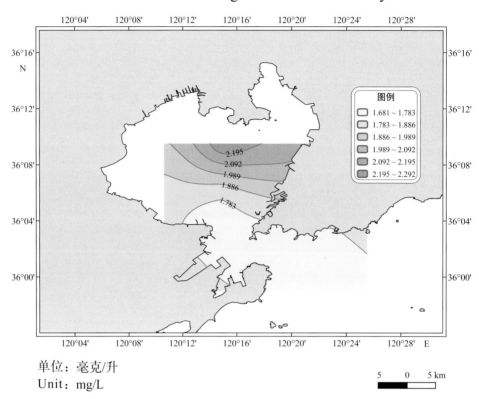

单位：毫克/升
Unit：mg/L

2012年5月溶解有机碳分布
Distribution of dissolved organic carbon in May 2012

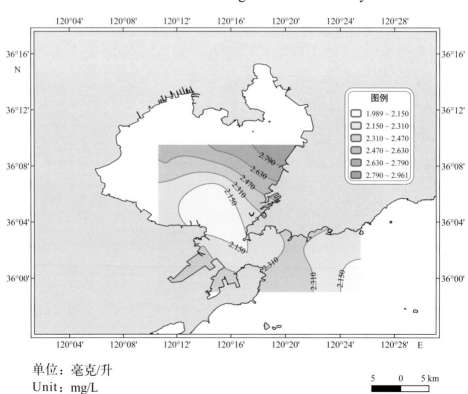

单位：毫克/升
Unit：mg/L

2012年8月溶解有机碳分布
Distribution of dissolved organic carbon in August 2012

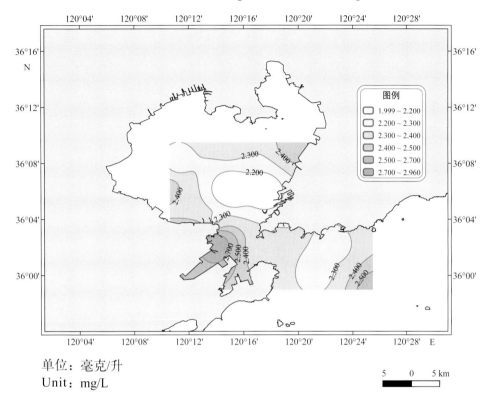

单位：毫克/升
Unit：mg/L

2012年11月溶解有机碳分布
Distribution of dissolved organic carbon in November 2012

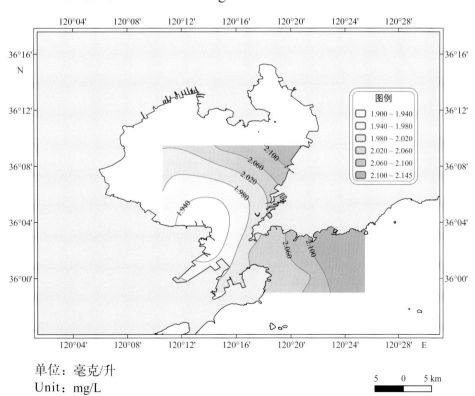

单位：毫克/升
Unit：mg/L

2013年2月溶解有机碳分布
Distribution of dissolved organic carbon in February 2013

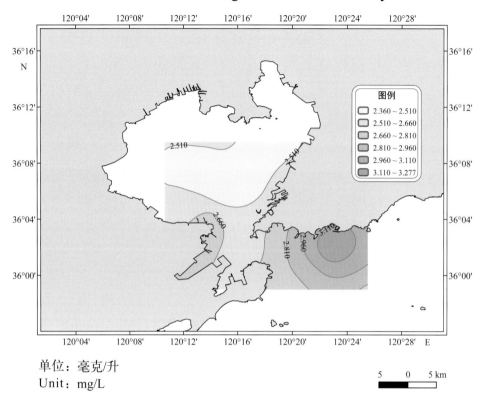

单位：毫克/升
Unit：mg/L

2013年5月溶解有机碳分布
Distribution of dissolved organic carbon in May 2013

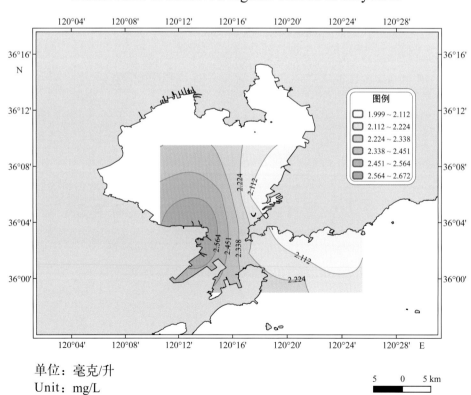

单位：毫克/升
Unit：mg/L

2013年8月溶解有机碳分布
Distribution of dissolved organic carbon in August 2013

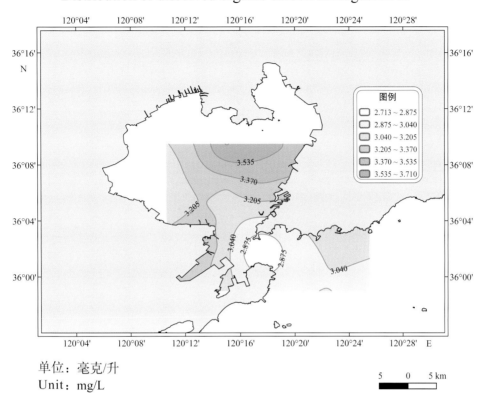

单位：毫克/升
Unit：mg/L

2013年11月溶解有机碳分布
Distribution of dissolved organic carbon in November 2013

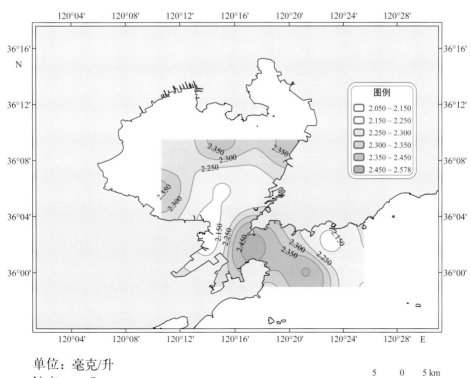

单位：毫克/升
Unit：mg/L

2014年2月溶解有机碳分布
Distribution of dissolved organic carbon in February 2014

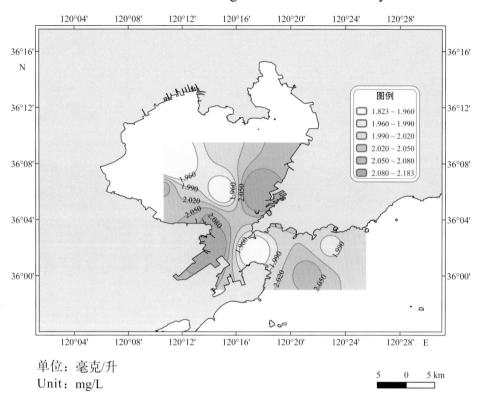

单位：毫克/升
Unit：mg/L

2014年5月溶解有机碳分布
Distribution of dissolved organic carbon in May 2014

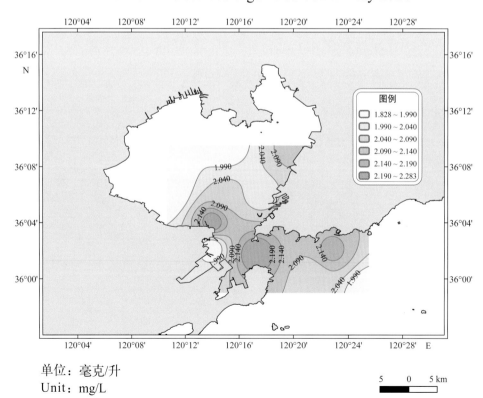

单位：毫克/升
Unit：mg/L

2014年8月溶解有机碳分布
Distribution of dissolved organic carbon in August 2014

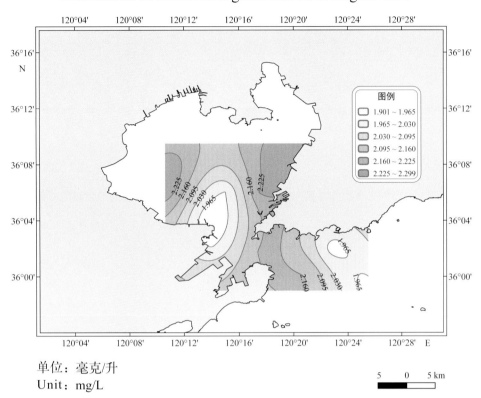

单位：毫克/升
Unit：mg/L

2014年11月溶解有机碳分布
Distribution of dissolved organic carbon in November 2014

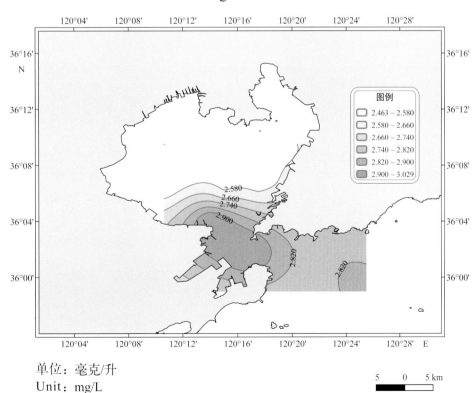

单位：毫克/升
Unit：mg/L

2015年2月溶解有机碳分布
Distribution of dissolved organic carbon in February 2015

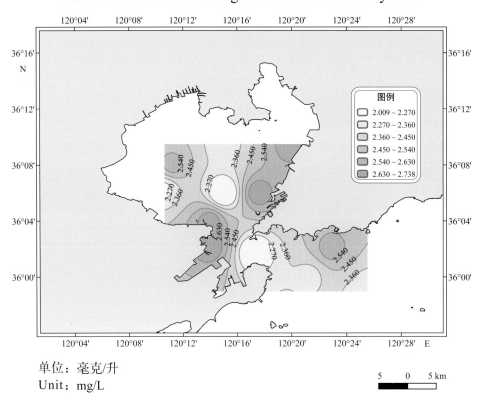

单位：毫克/升
Unit：mg/L

2015年5月溶解有机碳分布
Distribution of dissolved organic carbon in May 2015

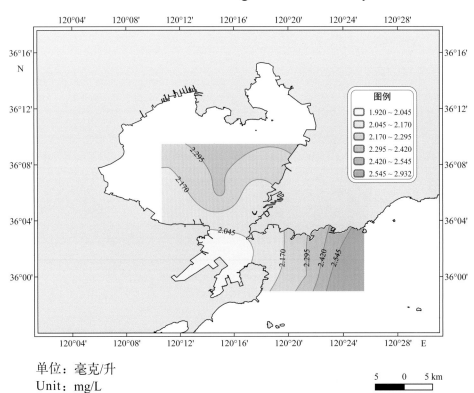

单位：毫克/升
Unit：mg/L

2015年8月溶解有机碳分布
Distribution of dissolved organic carbon in August 2015

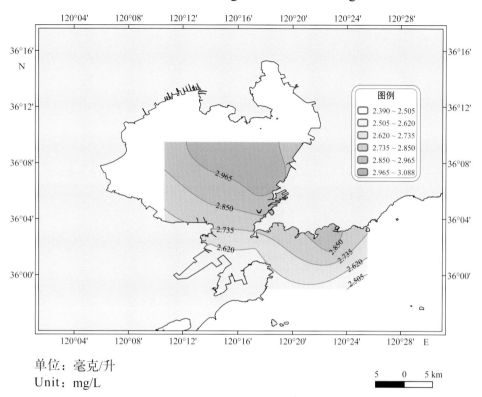

单位：毫克/升
Unit：mg/L

2015年11月溶解有机碳分布
Distribution of dissolved organic carbon in November 2015

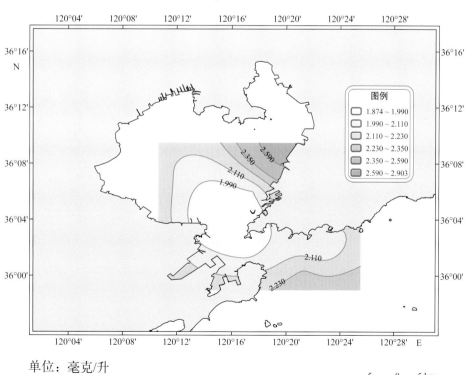

单位：毫克/升
Unit：mg/L

9. 化学需氧量 Chemical oxygen demand

2010年2月化学需氧量分布
Distribution of chemical oxygen demand in February 2010

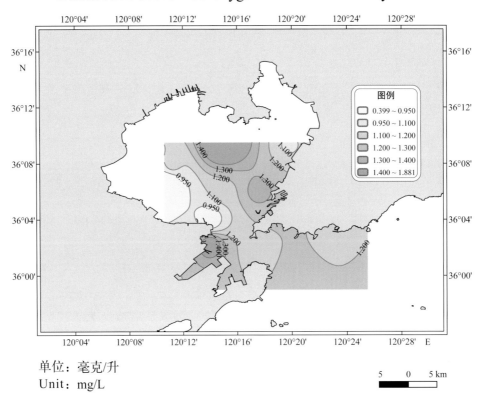

单位：毫克/升
Unit: mg/L

2010年5月化学需氧量分布
Distribution of chemical oxygen demand in May 2010

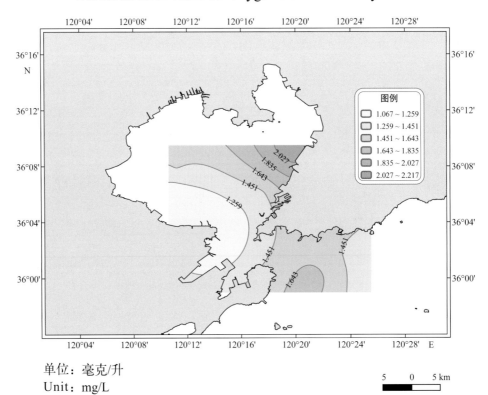

单位：毫克/升
Unit: mg/L

2010年8月化学需氧量分布
Distribution of chemical oxygen demand in August 2010

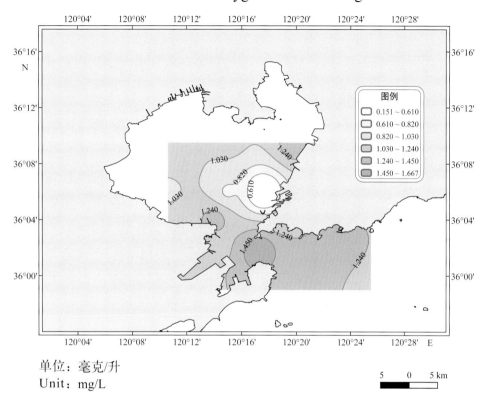

单位：毫克/升
Unit：mg/L

2010年11月化学需氧量分布
Distribution of chemical oxygen demand in November 2010

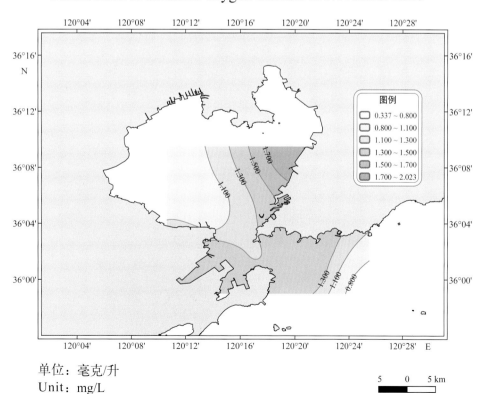

单位：毫克/升
Unit：mg/L

2011年2月化学需氧量分布
Distribution of chemical oxygen demand in February 2011

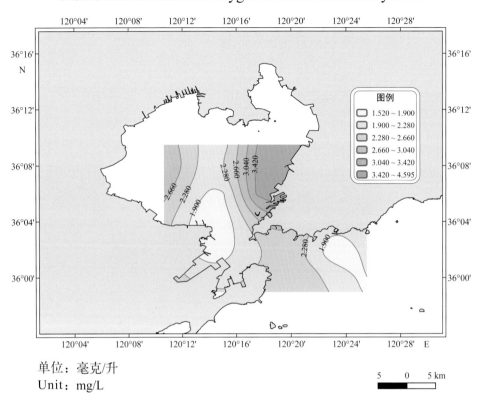

单位：毫克/升
Unit：mg/L

2011年5月化学需氧量分布
Distribution of chemical oxygen demand in May 2011

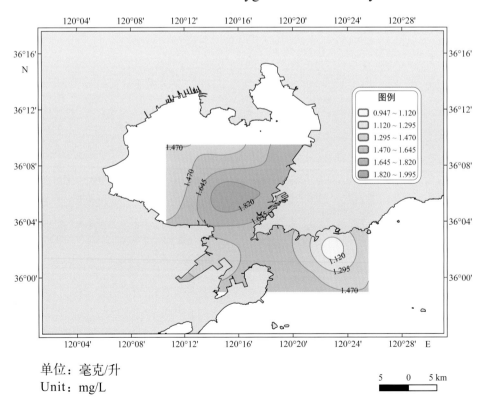

单位：毫克/升
Unit：mg/L

2011年8月化学需氧量分布
Distribution of chemical oxygen demand in August 2011

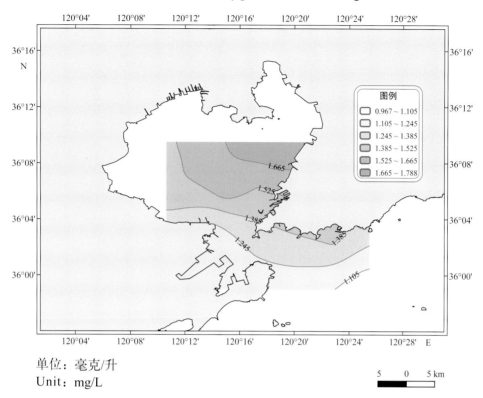

单位：毫克/升
Unit: mg/L

2011年11月化学需氧量分布
Distribution of chemical oxygen demand in November 2011

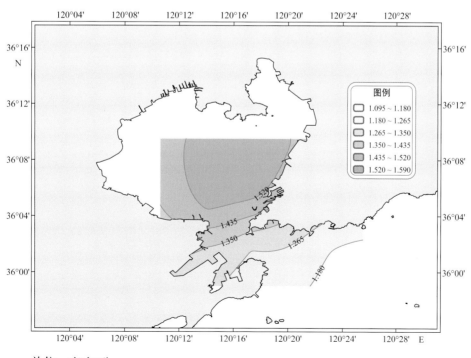

单位：毫克/升
Unit: mg/L

2012年2月化学需氧量分布
Distribution of chemical oxygen demand in February 2012

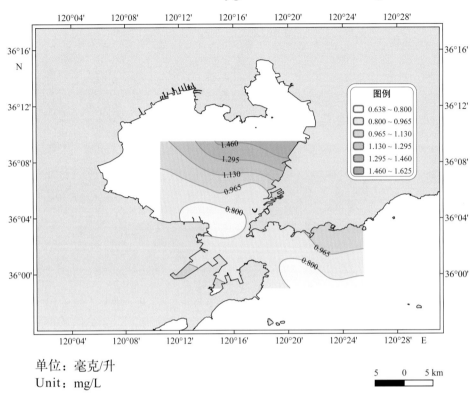

单位：毫克/升
Unit：mg/L

2012年5月化学需氧量分布
Distribution of chemical oxygen demand in May 2012

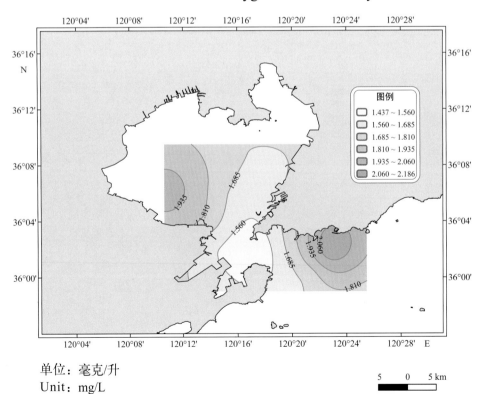

单位：毫克/升
Unit：mg/L

2012年8月化学需氧量分布
Distribution of chemical oxygen demand in August 2012

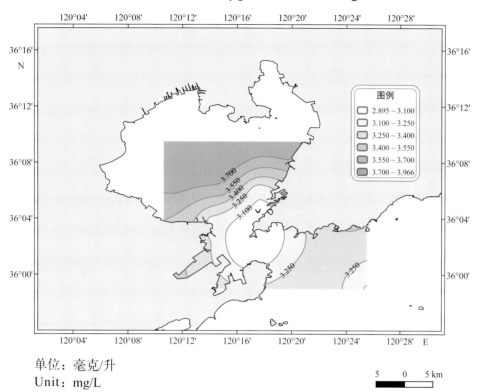

单位：毫克/升
Unit：mg/L

2012年11月化学需氧量分布
Distribution of chemical oxygen demand in November 2012

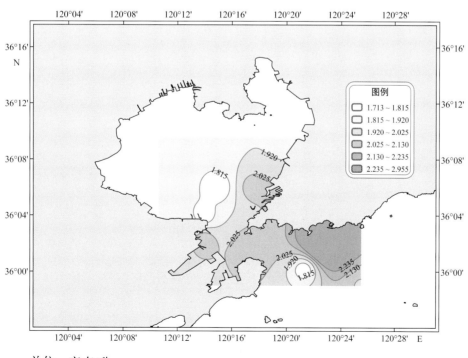

单位：毫克/升
Unit：mg/L

2013年2月化学需氧量分布
Distribution of chemical oxygen demand in February 2013

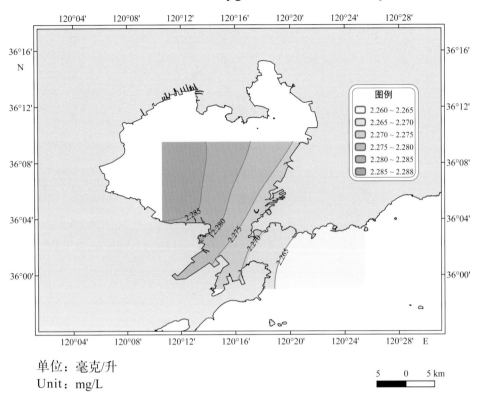

单位：毫克/升
Unit：mg/L

2013年5月化学需氧量分布
Distribution of chemical oxygen demand in May 2013

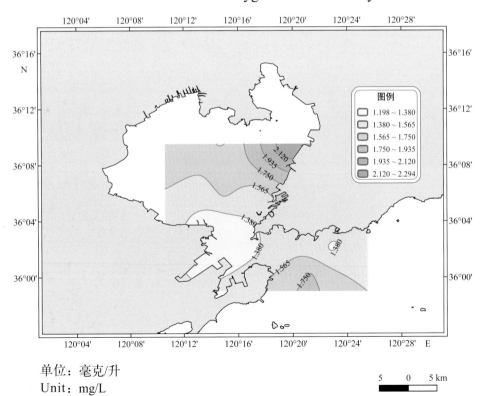

单位：毫克/升
Unit：mg/L

2013年8月化学需氧量分布
Distribution of chemical oxygen demand in August 2013

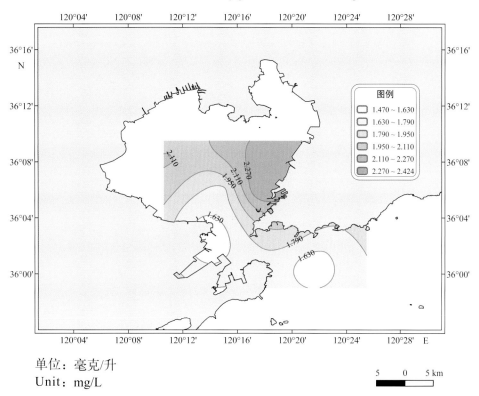

单位：毫克/升
Unit：mg/L

2013年11月化学需氧量分布
Distribution of chemical oxygen demand in November 2013

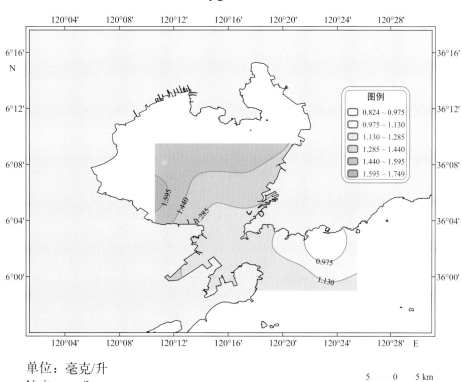

单位：毫克/升
Unit：mg/L

2014年2月化学需氧量分布
Distribution of chemical oxygen demand in February 2014

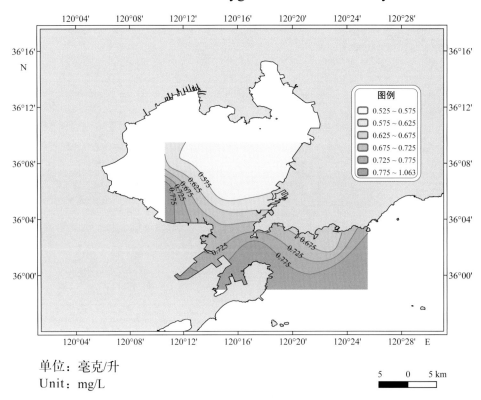

单位：毫克/升
Unit：mg/L

2014年5月化学需氧量分布
Distribution of chemical oxygen demand in May 2014

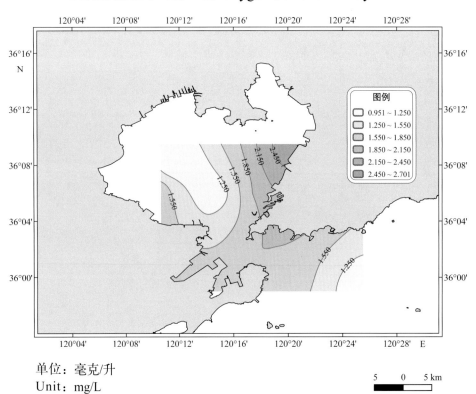

单位：毫克/升
Unit：mg/L

2014年8月化学需氧量分布
Distribution of chemical oxygen demand in August 2014

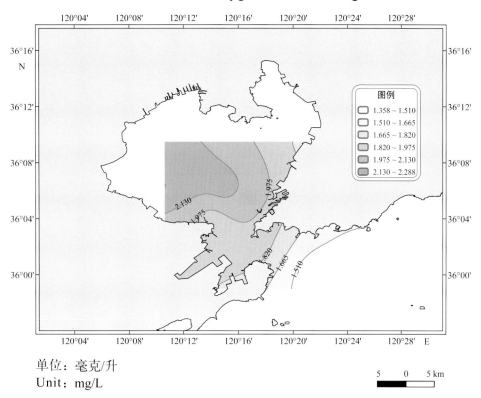

单位：毫克/升
Unit：mg/L

2014年11月化学需氧量分布
Distribution of chemical oxygen demand in November 2014

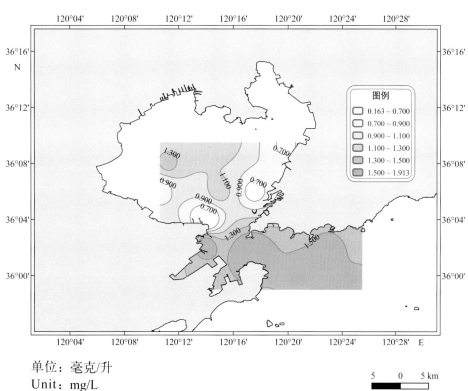

单位：毫克/升
Unit：mg/L

2015年2月化学需氧量分布
Distribution of chemical oxygen demand in February 2015

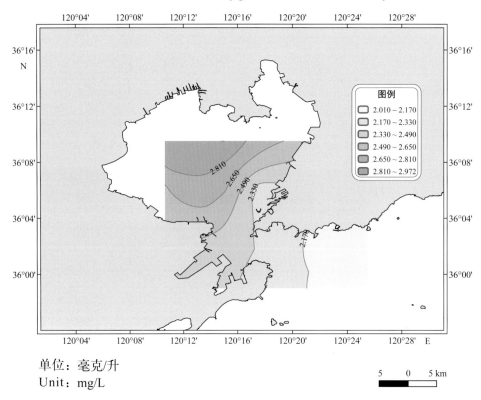

单位：毫克/升
Unit: mg/L

2015年5月化学需氧量分布
Distribution of chemical oxygen demand in May 2015

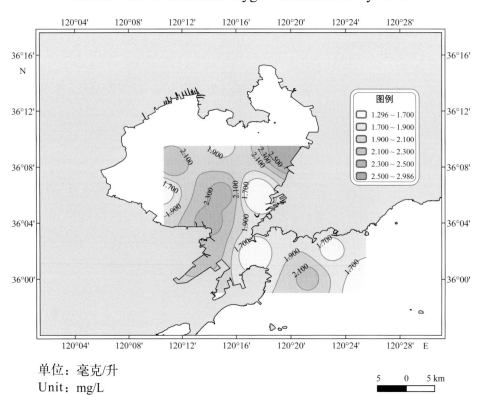

单位：毫克/升
Unit: mg/L

2015年8月化学需氧量分布
Distribution of chemical oxygen demand in August 2015

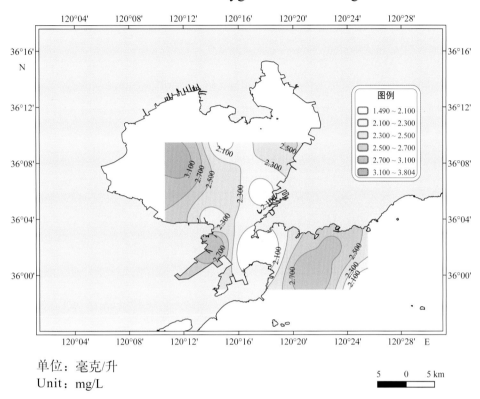

单位：毫克/升
Unit：mg/L

2015年11月化学需氧量分布
Distribution of chemical oxygen demand in November 2015

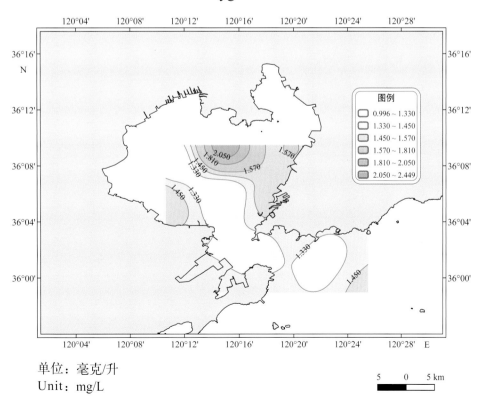

单位：毫克/升
Unit：mg/L

10. 总磷 Total phosphorus

2010年2月总磷分布
Distribution of total phosphorus in February 2010

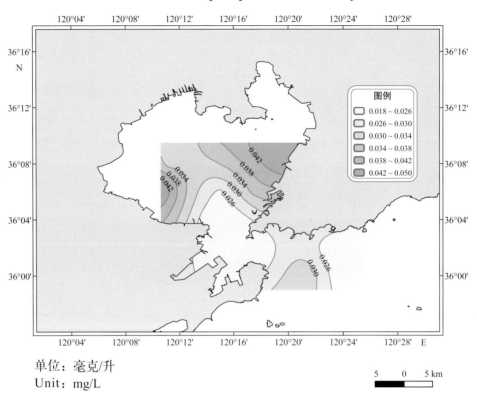

单位：毫克/升
Unit：mg/L

2010年5月总磷分布
Distribution of total phosphorus in May 2010

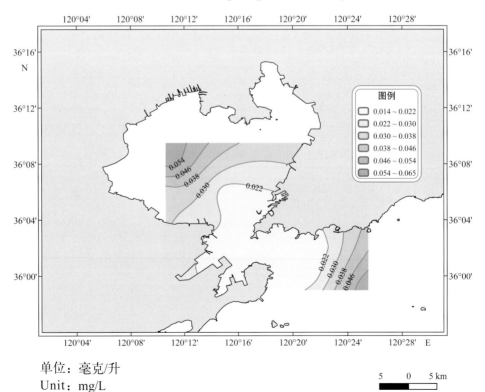

单位：毫克/升
Unit：mg/L

2010年8月总磷分布
Distribution of total phosphorus in August 2010

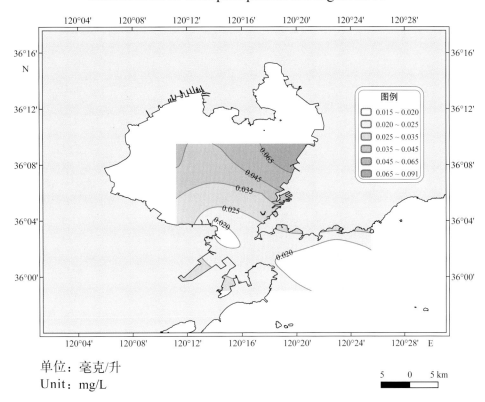

单位：毫克/升
Unit：mg/L

2010年11月总磷分布
Distribution of total phosphorus in November 2010

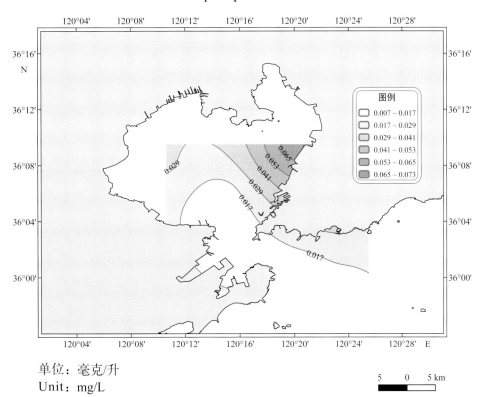

单位：毫克/升
Unit：mg/L

2011年2月总磷分布
Distribution of total phosphorus in February 2011

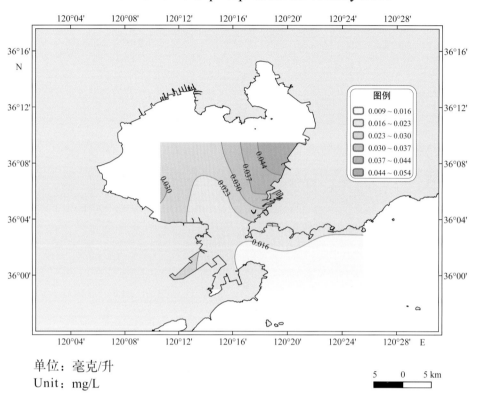

单位：毫克/升
Unit：mg/L

2011年5月总磷分布
Distribution of total phosphorus in May 2011

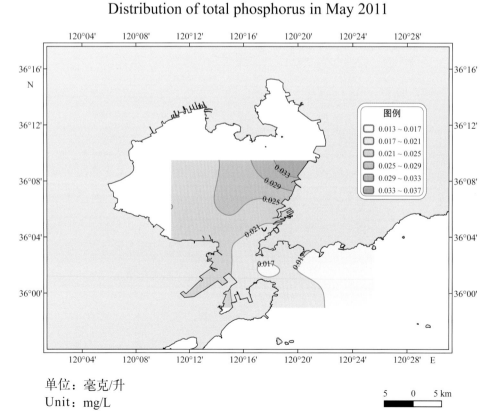

单位：毫克/升
Unit：mg/L

2011年8月总磷分布
Distribution of total phosphorus in August 2011

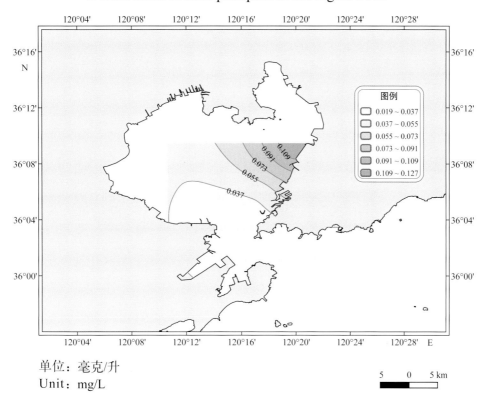

单位：毫克/升
Unit：mg/L

2011年11月总磷分布
Distribution of total phosphorus in November 2011

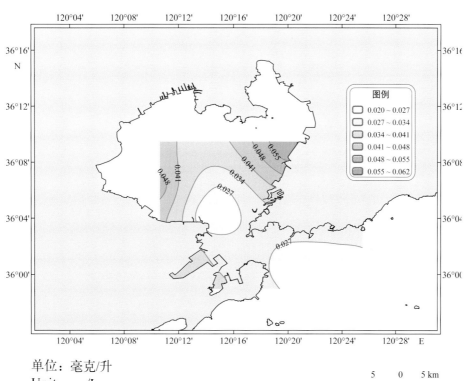

单位：毫克/升
Unit：mg/L

2012年2月总磷分布
Distribution of total phosphorus in February 2012

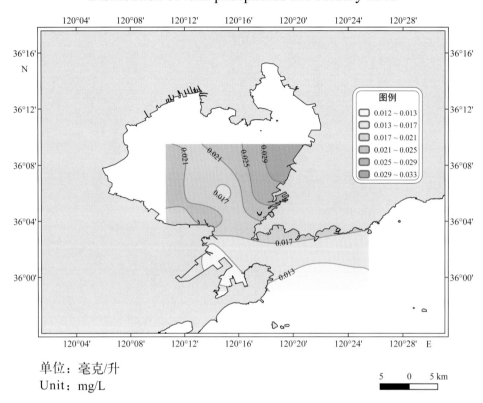

单位：毫克/升
Unit：mg/L

2012年5月总磷分布
Distribution of total phosphorus in May 2012

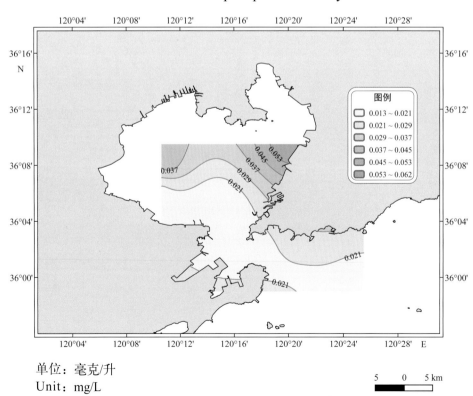

单位：毫克/升
Unit：mg/L

2012年8月总磷分布
Distribution of total phosphorus in August 2012

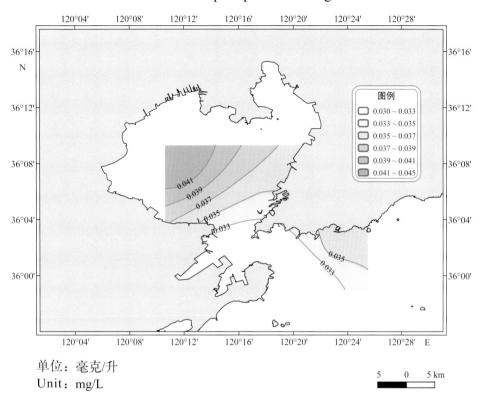

单位：毫克/升
Unit：mg/L

2012年11月总磷分布
Distribution of total phosphorus in November 2012

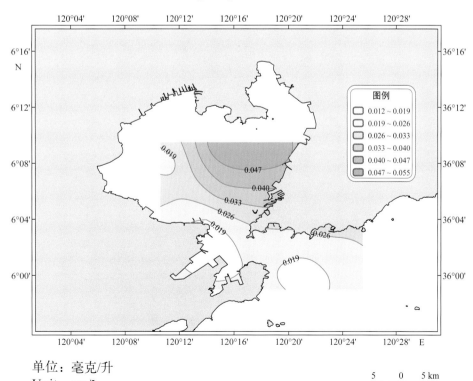

单位：毫克/升
Unit：mg/L

2013年2月总磷分布
Distribution of total phosphorus in February 2013

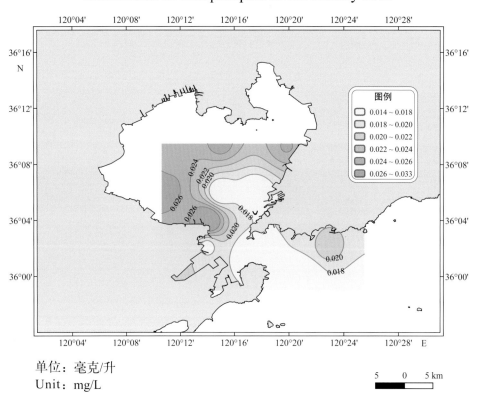

单位：毫克/升
Unit：mg/L

2013年5月总磷分布
Distribution of total phosphorus in May 2013

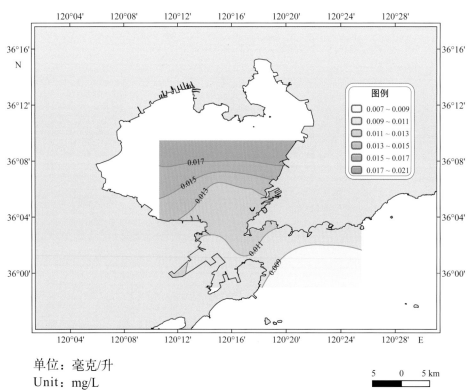

单位：毫克/升
Unit：mg/L

2013年8月总磷分布
Distribution of total phosphorus in August 2013

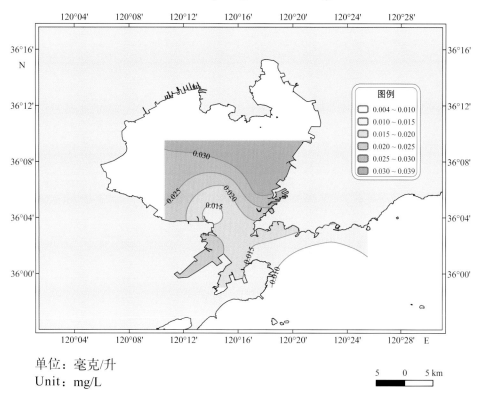

单位：毫克/升
Unit：mg/L

2013年11月总磷分布
Distribution of total phosphorus in November 2013

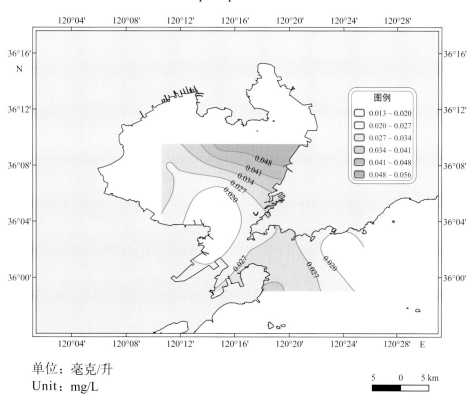

单位：毫克/升
Unit：mg/L

2014年2月总磷分布
Distribution of total phosphorus in February 2014

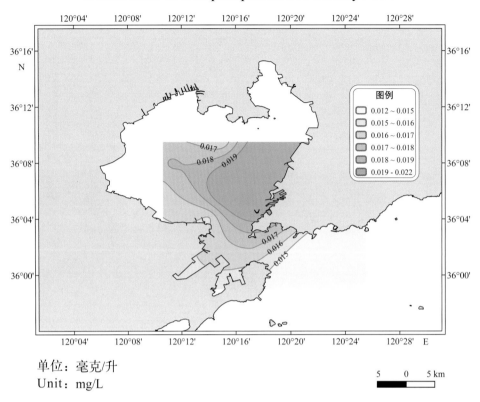

单位：毫克/升
Unit：mg/L

2014年5月总磷分布
Distribution of total phosphorus in May 2014

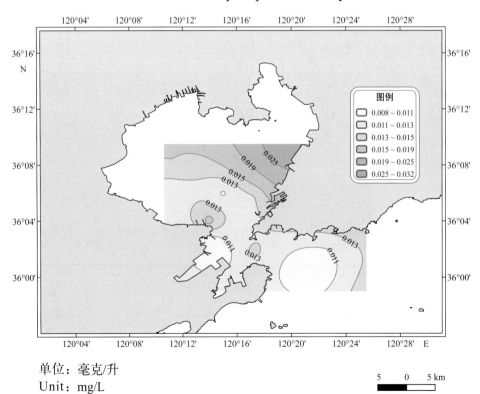

单位：毫克/升
Unit：mg/L

2014年8月总磷分布
Distribution of total phosphorus in August 2014

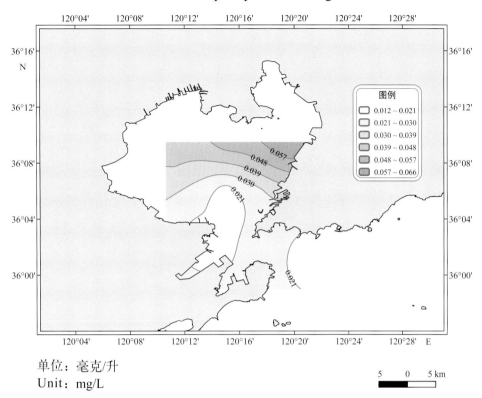

单位：毫克/升
Unit：mg/L

2014年11月总磷分布
Distribution of total phosphorus in November 2014

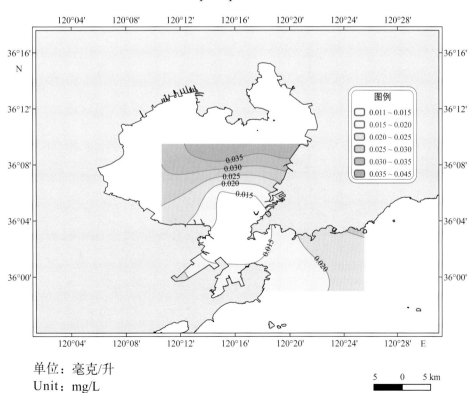

单位：毫克/升
Unit：mg/L

2015年2月总磷分布
Distribution of total phosphorus in February 2015

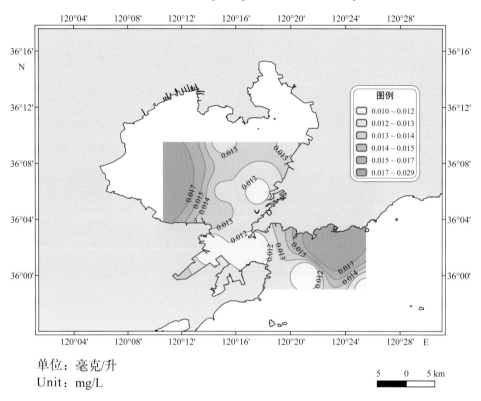

单位：毫克/升
Unit：mg/L

2015年5月总磷分布
Distribution of total phosphorus in May 2015

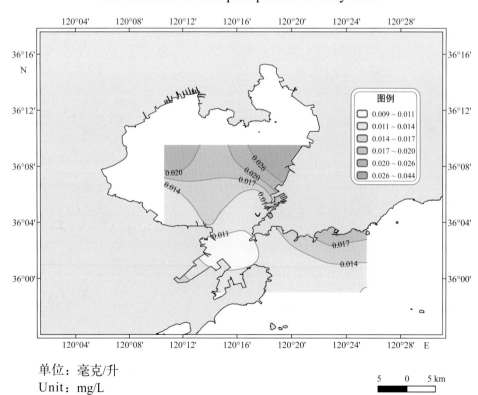

单位：毫克/升
Unit：mg/L

2015年8月总磷分布
Distribution of total phosphorus in August 2015

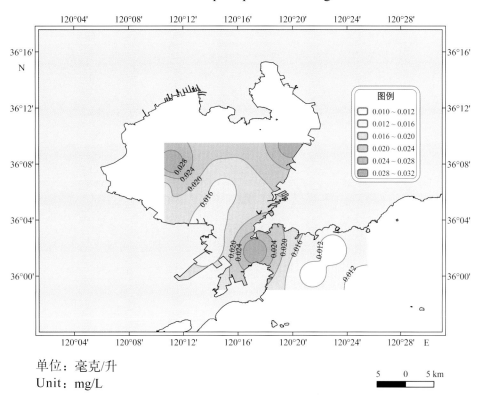

单位：毫克/升
Unit：mg/L

2015年11月总磷分布
Distribution of total phosphorus in November 2015

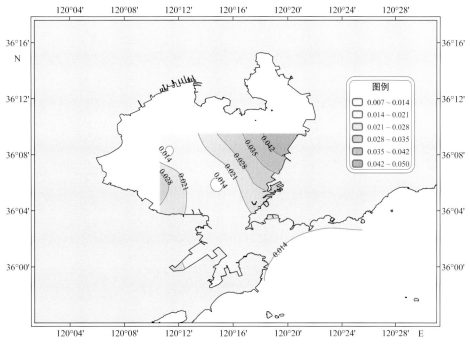

单位：毫克/升
Unit：mg/L

11. 总氮 Total nitrogen

2010年2月总氮分布
Distribution of total nitrogen in February 2010

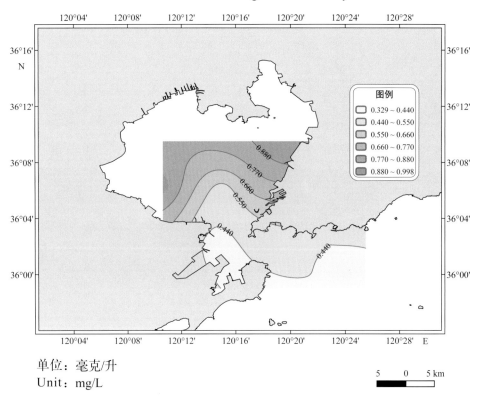

单位：毫克/升
Unit：mg/L

2010年5月总氮分布
Distribution of total nitrogen in May 2010

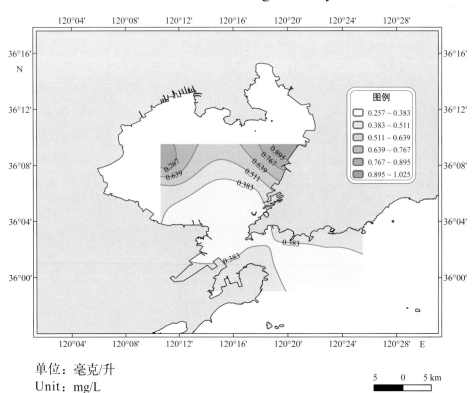

单位：毫克/升
Unit：mg/L

2010年8月总氮分布
Distribution of total nitrogen in August 2010

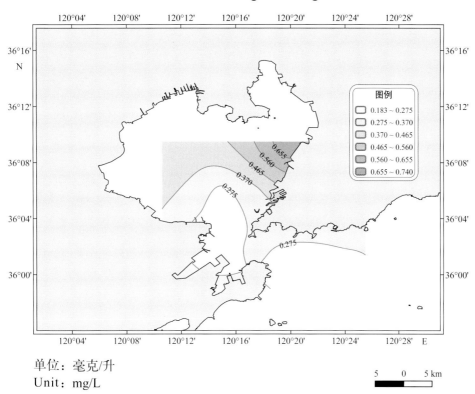

单位：毫克/升
Unit：mg/L

2010年11月总氮分布
Distribution of total nitrogen in November 2010

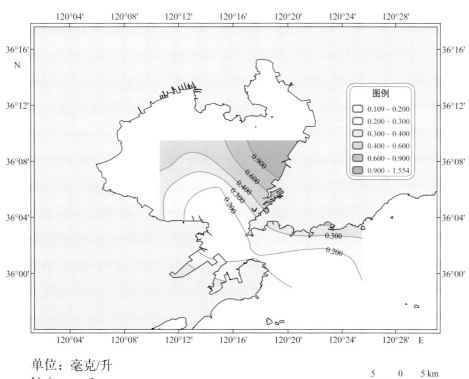

单位：毫克/升
Unit：mg/L

2011年2月总氮分布
Distribution of total nitrogen in February 2011

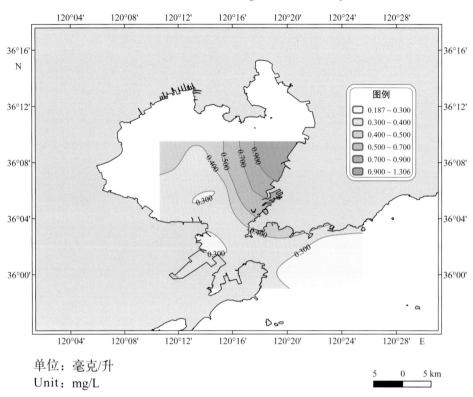

单位：毫克/升
Unit：mg/L

2011年5月总氮分布
Distribution of total nitrogen in May 2011

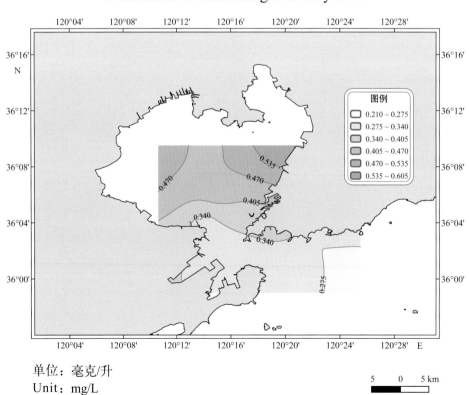

单位：毫克/升
Unit：mg/L

2011年8月总氮分布
Distribution of total nitrogen in August 2011

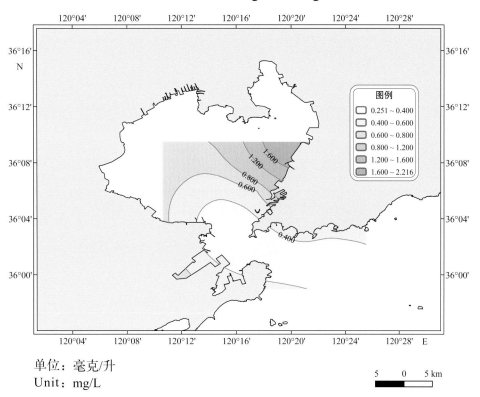

单位：毫克/升
Unit：mg/L

2011年11月总氮分布
Distribution of total nitrogen in November 2011

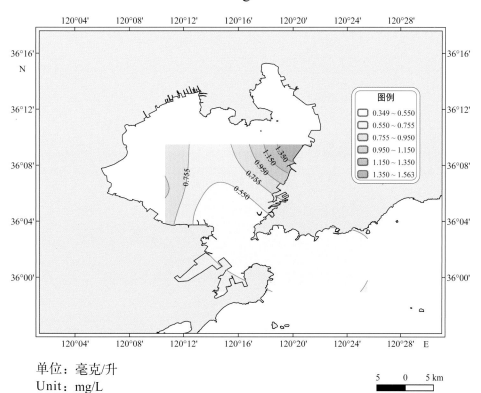

单位：毫克/升
Unit：mg/L

2012年2月总氮分布
Distribution of total nitrogen in February 2012

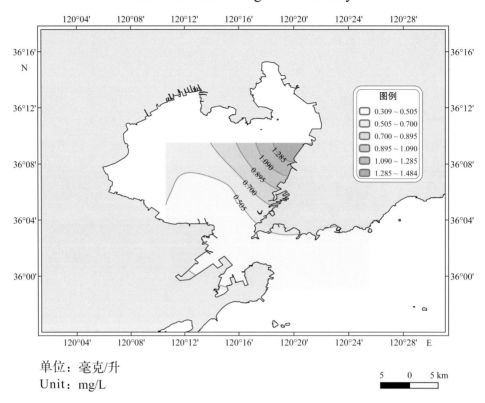

单位：毫克/升
Unit：mg/L

2012年5月总氮分布
Distribution of total nitrogen in May 2012

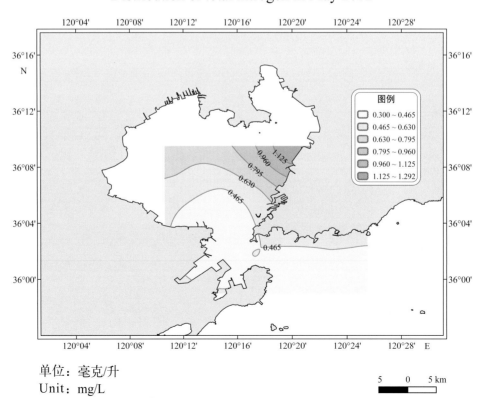

单位：毫克/升
Unit：mg/L

2012年8月总氮分布
Distribution of total nitrogen in August 2012

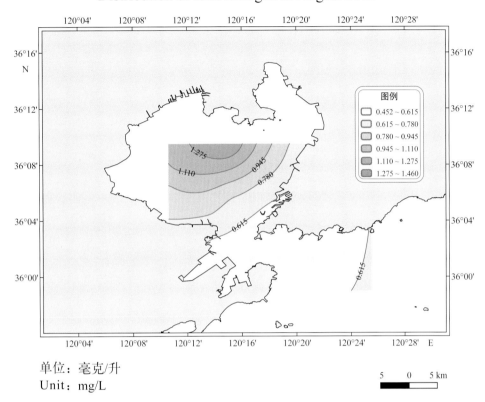

单位：毫克/升
Unit：mg/L

2012年11月总氮分布
Distribution of total nitrogen in November 2012

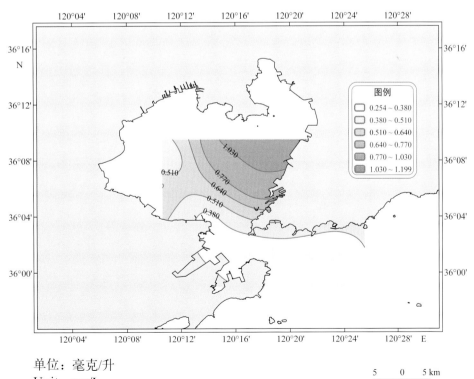

单位：毫克/升
Unit：mg/L

2013年2月总氮分布
Distribution of total nitrogen in February 2013

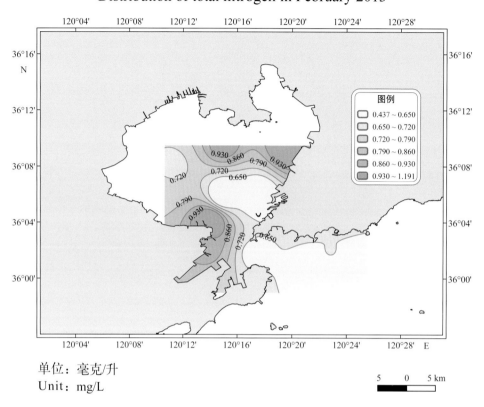

单位：毫克/升
Unit：mg/L

2013年5月总氮分布
Distribution of total nitrogen in May 2013

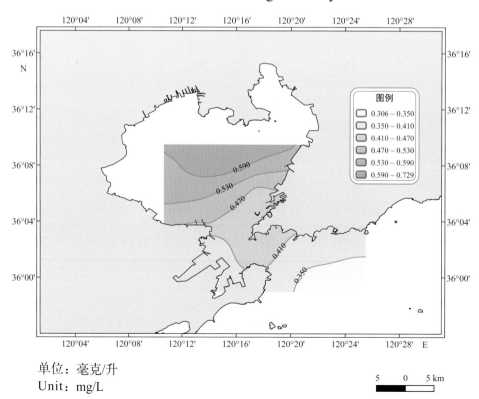

单位：毫克/升
Unit：mg/L

2013年8月总氮分布
Distribution of total nitrogen in August 2013

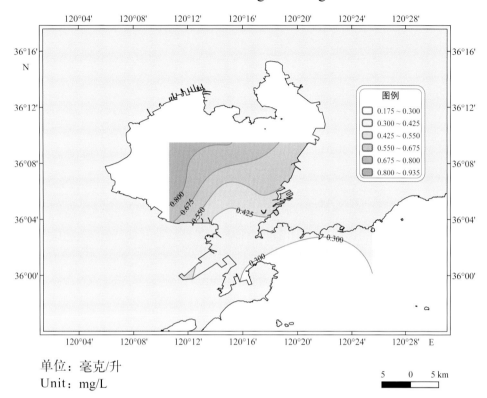

单位：毫克/升
Unit：mg/L

2013年11月总氮分布
Distribution of total nitrogen in November 2013

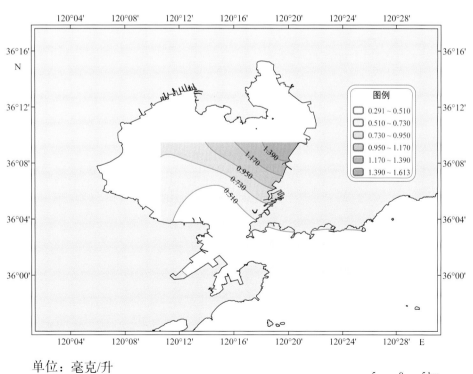

单位：毫克/升
Unit：mg/L

2014年2月总氮分布
Distribution of total nitrogen in February 2014

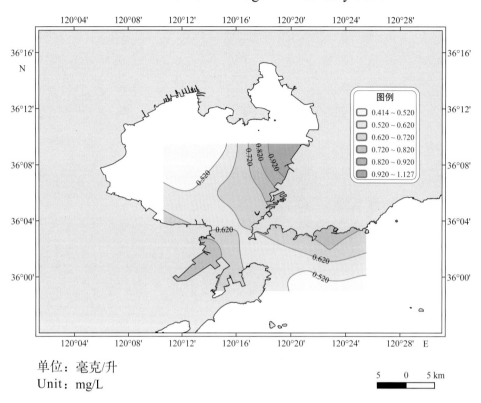

单位：毫克/升
Unit：mg/L

2014年5月总氮分布
Distribution of total nitrogen in May 2014

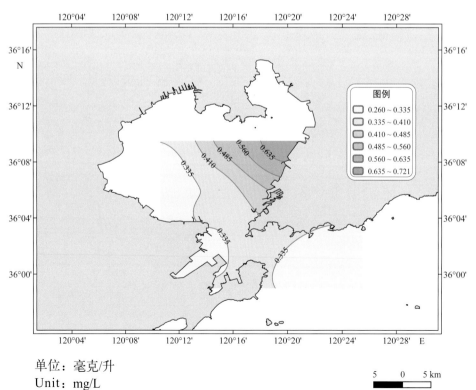

单位：毫克/升
Unit：mg/L

2014年8月总氮分布
Distribution of total nitrogen in August 2014

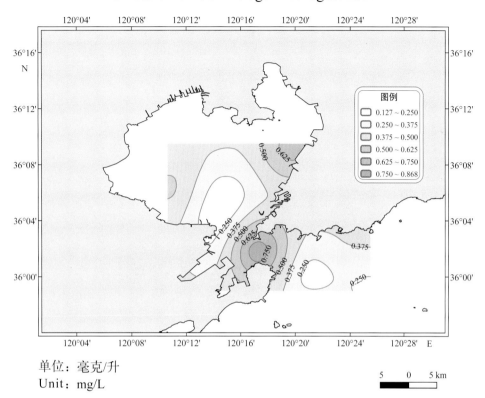

单位：毫克/升
Unit：mg/L

2014年11月总氮分布
Distribution of total nitrogen in November 2014

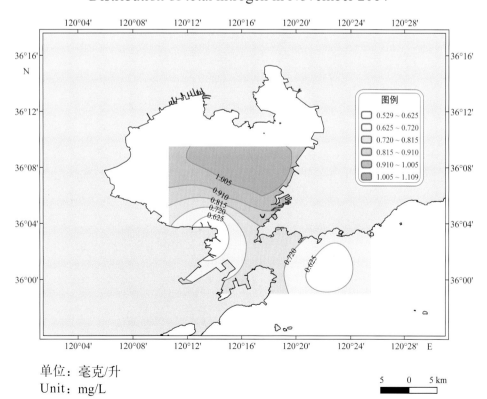

单位：毫克/升
Unit：mg/L

2015年2月总氮分布
Distribution of total nitrogen in February 2015

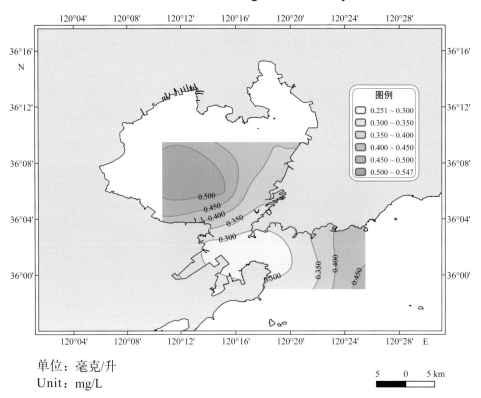

单位：毫克/升
Unit: mg/L

2015年5月总氮分布
Distribution of total nitrogen in May 2015

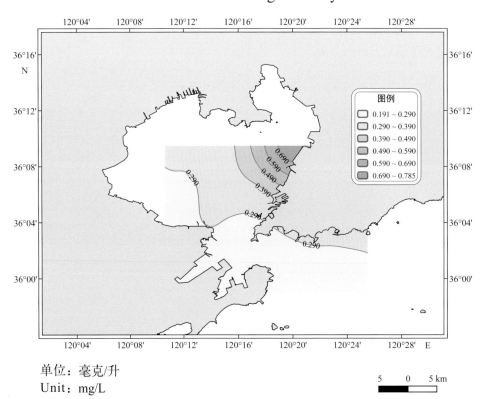

单位：毫克/升
Unit: mg/L

2015年8月总氮分布
Distribution of total nitrogen in August 2015

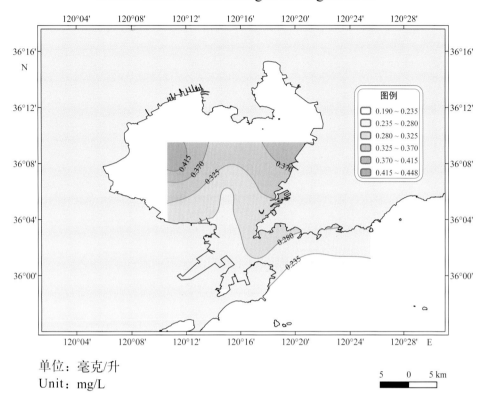

单位：毫克/升
Unit：mg/L

2015年11月总氮分布
Distribution of total nitrogen in November 2015

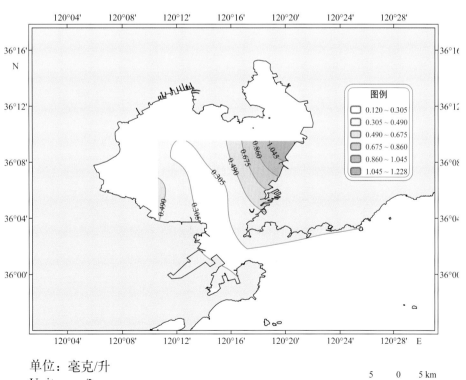

单位：毫克/升
Unit：mg/L

12. 颗粒有机碳 Particulate organic carbon

2010年2月颗粒有机碳分布
Distribution of particulate organic carbon in February 2010

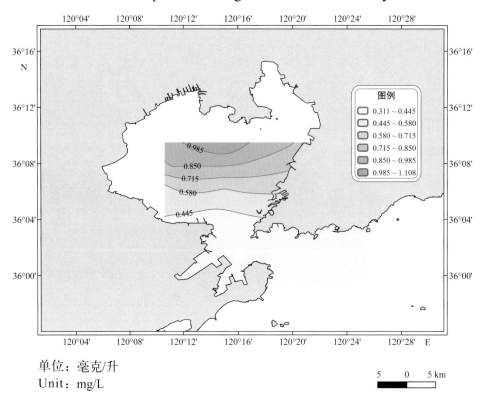

单位：毫克/升
Unit：mg/L

2010年5月颗粒有机碳分布
Distribution of particulate organic carbon in May 2010

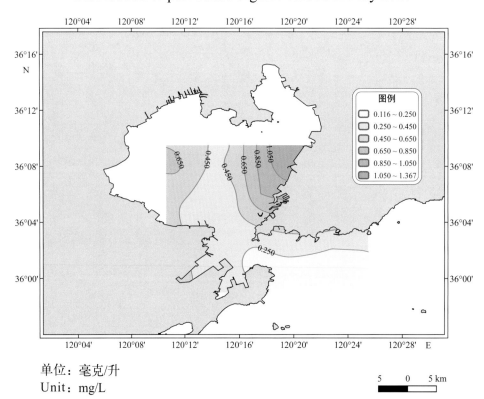

单位：毫克/升
Unit：mg/L

2010年8月颗粒有机碳分布
Distribution of particulate organic carbon in August 2010

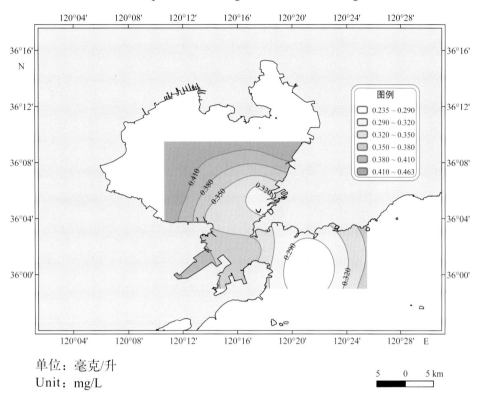

单位：毫克/升
Unit：mg/L

2011年2月颗粒有机碳分布
Distribution of particulate organic carbon in February 2011

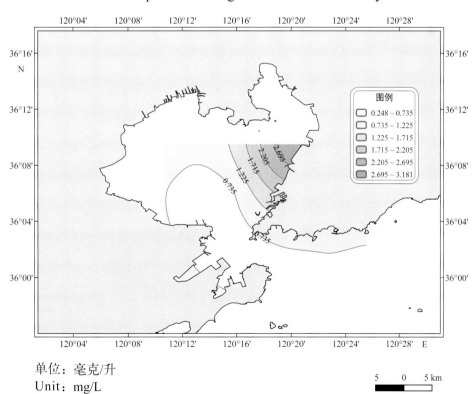

单位：毫克/升
Unit：mg/L

2011年5月颗粒有机碳分布
Distribution of particulate organic carbon in May 2011

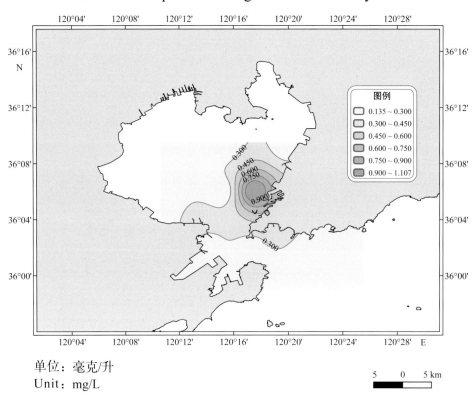

单位：毫克/升
Unit：mg/L

2011年8月颗粒有机碳分布
Distribution of particulate organic carbon in August 2011

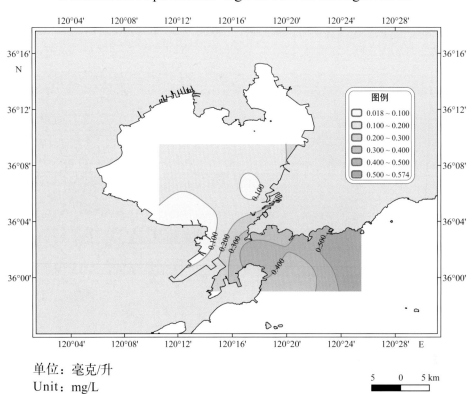

单位：毫克/升
Unit：mg/L

2011年11月颗粒有机碳分布
Distribution of particulate organic carbon in November 2011

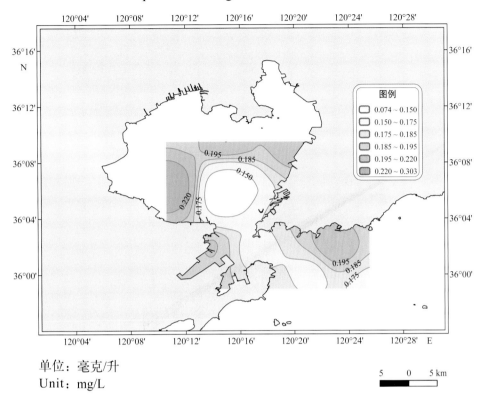

单位：毫克/升
Unit：mg/L

2012年2月颗粒有机碳分布
Distribution of particulate organic carbon in February 2012

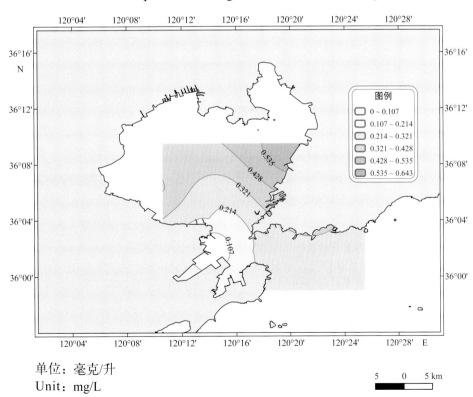

单位：毫克/升
Unit：mg/L

2012年5月颗粒有机碳分布
Distribution of particulate organic carbon in May 2012

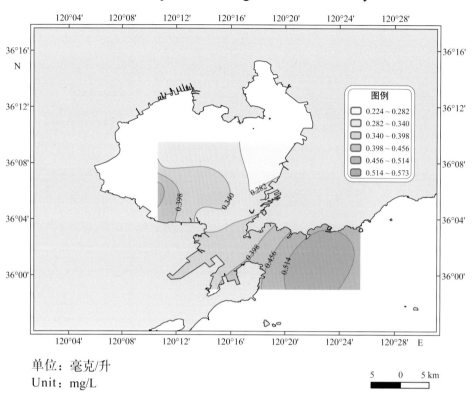

单位：毫克/升
Unit：mg/L

2012年8月颗粒有机碳分布
Distribution of particulate organic carbon in August 2012

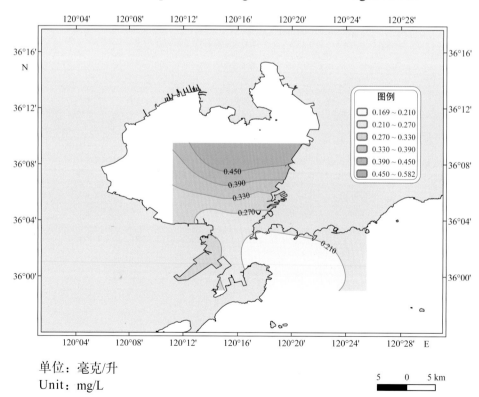

单位：毫克/升
Unit：mg/L

2012年11月颗粒有机碳分布
Distribution of particulate organic carbon in November 2012

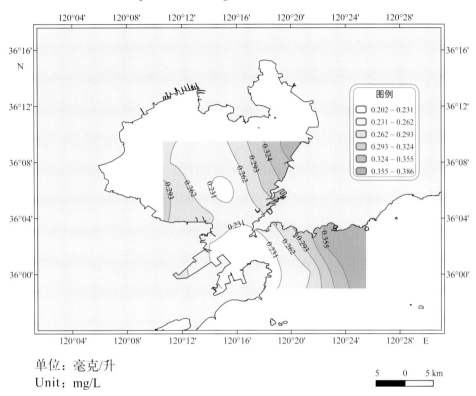

单位：毫克/升
Unit: mg/L

2013年2月颗粒有机碳分布
Distribution of particulate organic carbon in February 2013

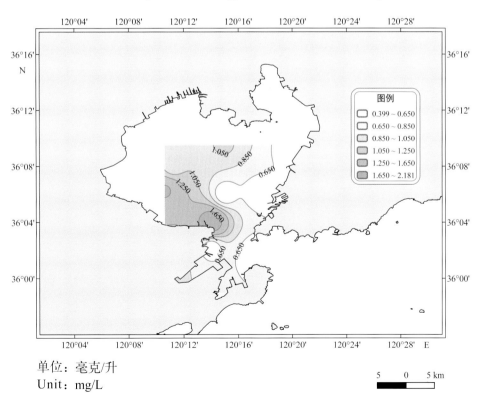

单位：毫克/升
Unit: mg/L

2013年5月颗粒有机碳分布
Distribution of particulate organic carbon in May 2013

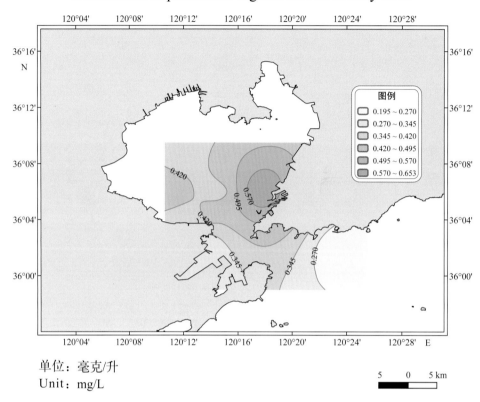

单位：毫克/升
Unit: mg/L

2013年8月颗粒有机碳分布
Distribution of particulate organic carbon in August 2013

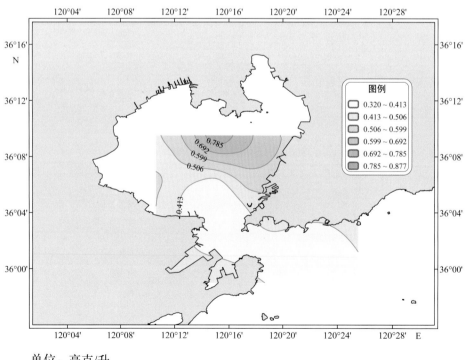

单位：毫克/升
Unit: mg/L

2013年11月颗粒有机碳分布
Distribution of particulate organic carbon in November 2013

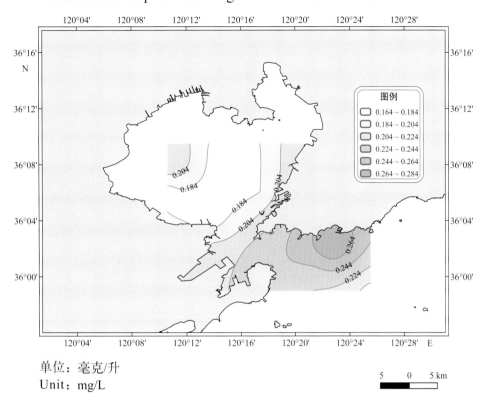

单位：毫克/升
Unit：mg/L

2014年2月颗粒有机碳分布
Distribution of particulate organic carbon in February 2014

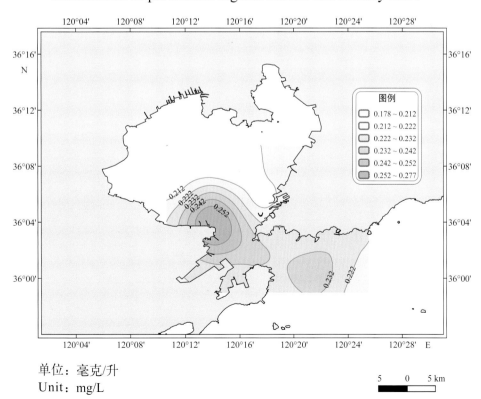

单位：毫克/升
Unit：mg/L

2014年5月颗粒有机碳分布
Distribution of particulate organic carbon in May 2014

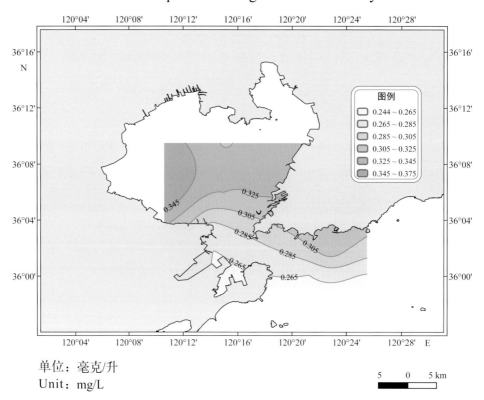

单位：毫克/升
Unit：mg/L

2014年8月颗粒有机碳分布
Distribution of particulate organic carbon in August 2014

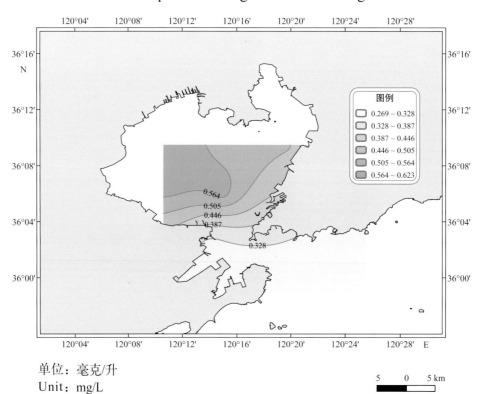

单位：毫克/升
Unit：mg/L

2014年11月颗粒有机碳分布
Distribution of particulate organic carbon in November 2014

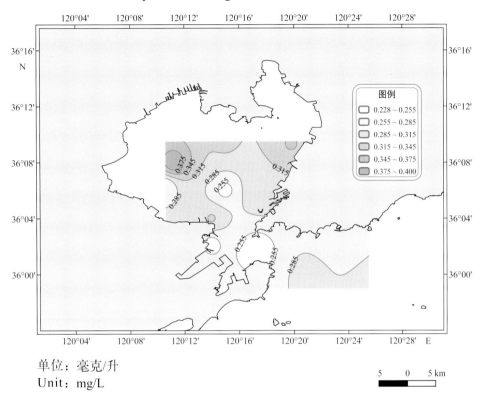

单位：毫克/升
Unit：mg/L

2015年2月颗粒有机碳分布
Distribution of particulate organic carbon in February 2015

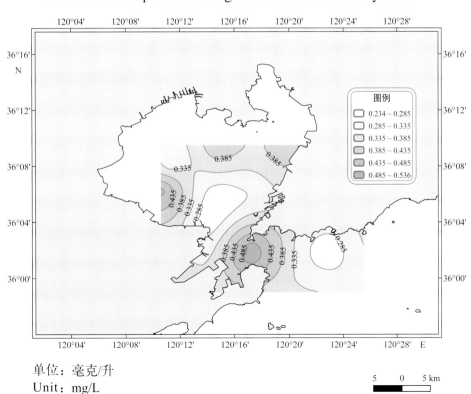

单位：毫克/升
Unit：mg/L

2015年5月颗粒有机碳分布
Distribution of particulate organic carbon in May 2015

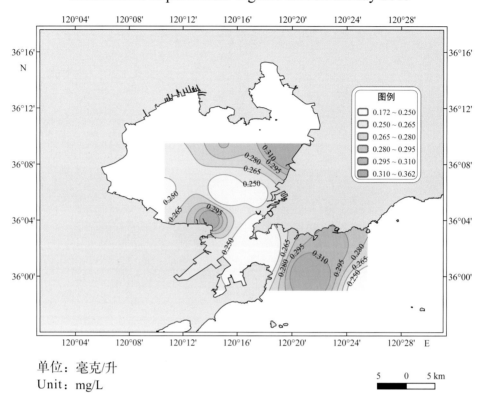

单位：毫克/升
Unit：mg/L

2015年8月颗粒有机碳分布
Distribution of particulate organic carbon in August 2015

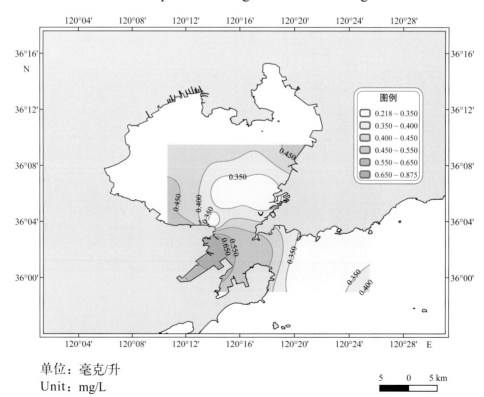

单位：毫克/升
Unit：mg/L

2015年11月颗粒有机碳分布
Distribution of particulate organic carbon in November 2015

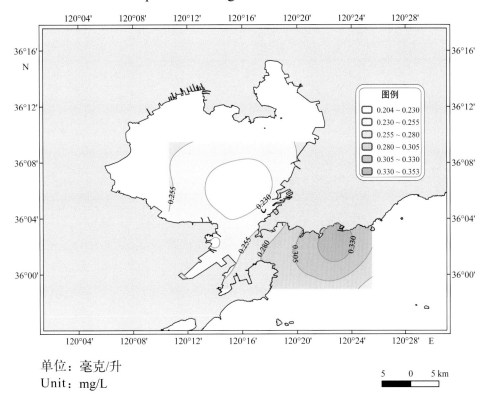

单位：毫克/升
Unit：mg/L

三、悬浮体和沉积物

SUSPENDED MATTER AND SEDIMENT

1. 悬浮体浓度 Suspended matter concentration

2010年2月悬浮体浓度分布
Distribution of suspended matter concentration in February 2010

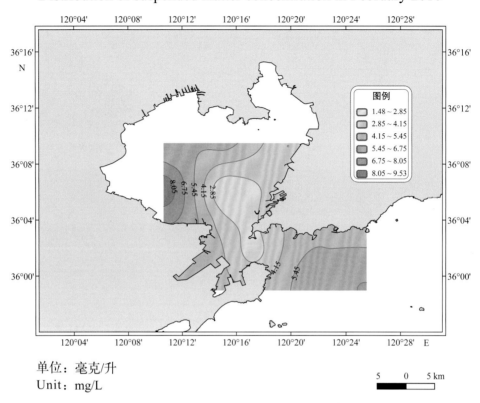

单位：毫克/升
Unit：mg/L

2010年5月悬浮体浓度分布
Distribution of suspended matter concentration in May 2010

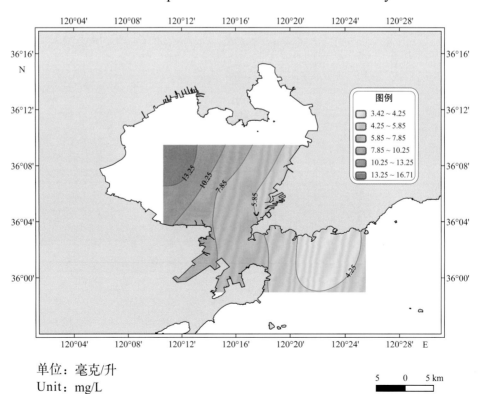

单位：毫克/升
Unit：mg/L

2010年8月悬浮体浓度分布
Distribution of suspended matter concentration in August 2010

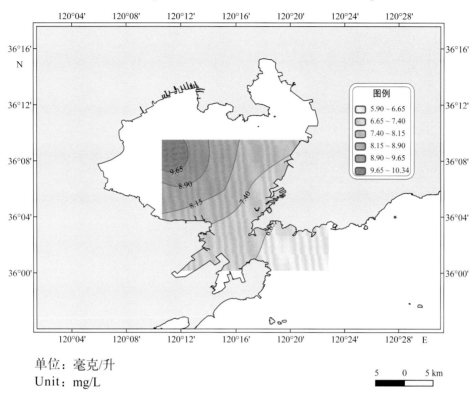

单位：毫克/升
Unit：mg/L

2010年11月悬浮体浓度分布
Distribution of suspended matter concentration in November 2010

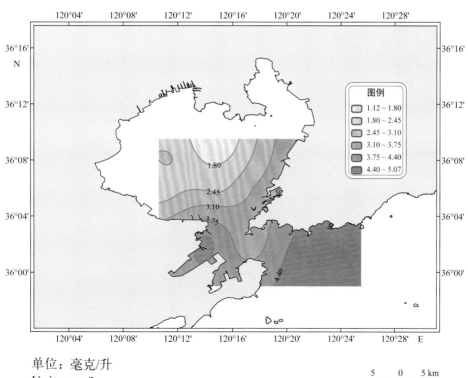

单位：毫克/升
Unit：mg/L

2011年2月悬浮体浓度分布
Distribution of suspended matter concentration in February 2011

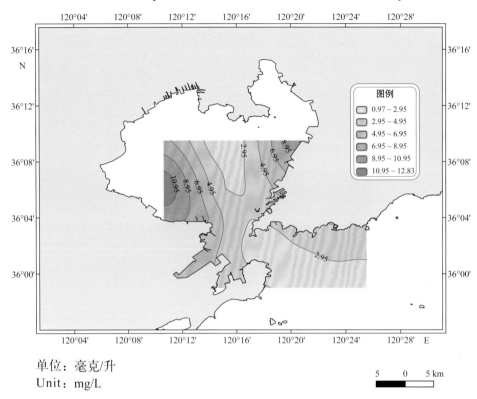

单位：毫克/升
Unit：mg/L

2011年5月悬浮体浓度分布
Distribution of suspended matter concentration in May 2011

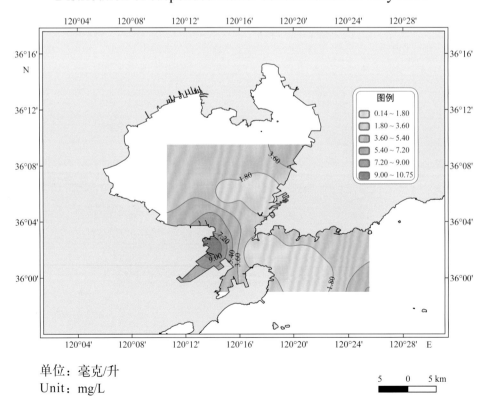

单位：毫克/升
Unit：mg/L

2011年8月悬浮体浓度分布
Distribution of suspended matter concentration in August 2011

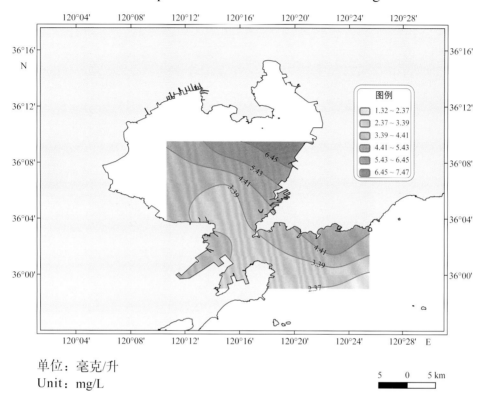

单位：毫克/升
Unit：mg/L

2011年11月悬浮体浓度分布
Distribution of suspended matter concentration in November 2011

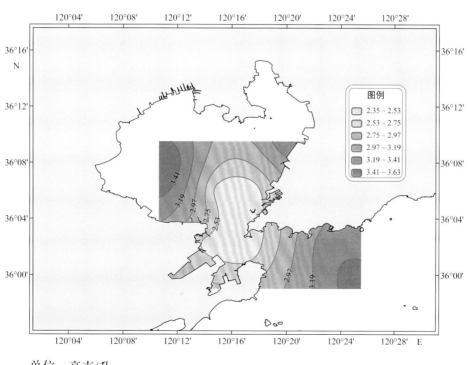

单位：毫克/升
Unit：mg/L

2012年2月悬浮体浓度分布
Distribution of suspended matter concentration in February 2012

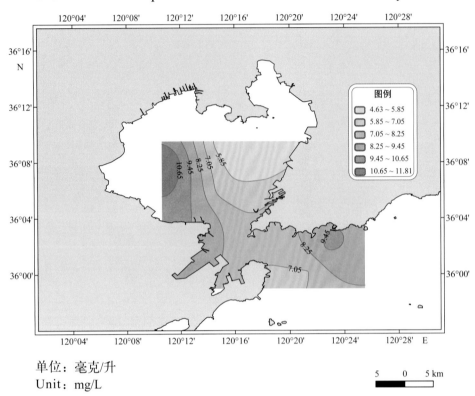

单位：毫克/升
Unit：mg/L

2012年5月悬浮体浓度分布
Distribution of suspended matter concentration in May 2012

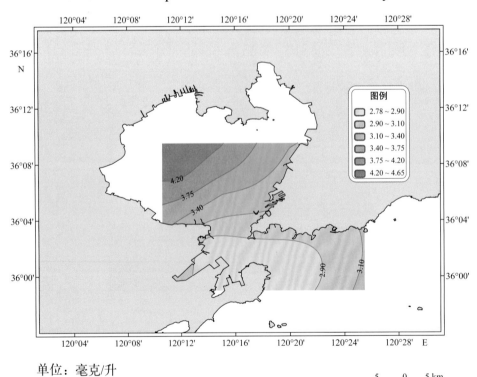

单位：毫克/升
Unit：mg/L

2012年8月悬浮体浓度分布
Distribution of suspended matter concentration in August 2012

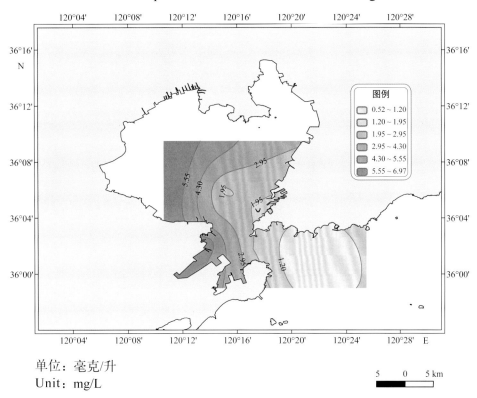

单位：毫克/升
Unit：mg/L

2012年11月悬浮体浓度分布
Distribution of suspended matter concentration in November 2012

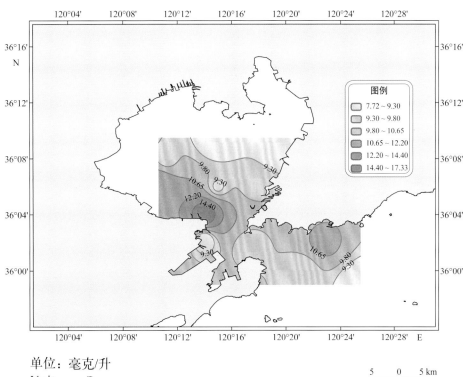

单位：毫克/升
Unit：mg/L

2013年2月悬浮体浓度分布
Distribution of suspended matter concentration in February 2013

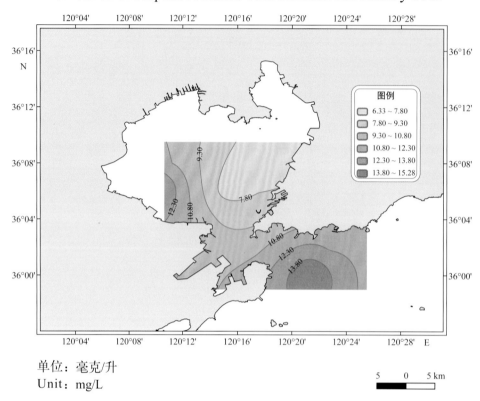

单位:毫克/升
Unit: mg/L

2013年5月悬浮体浓度分布
Distribution of suspended matter concentration in May 2013

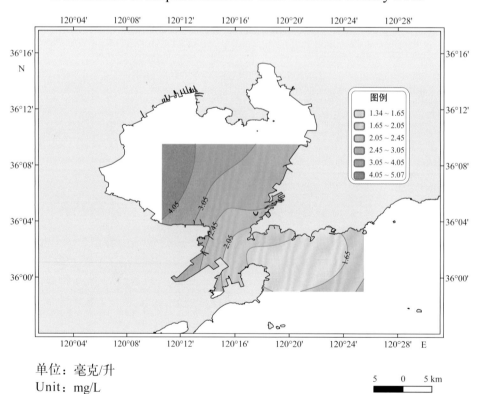

单位:毫克/升
Unit: mg/L

2013年8月悬浮体浓度分布
Distribution of suspended matter concentration in August 2013

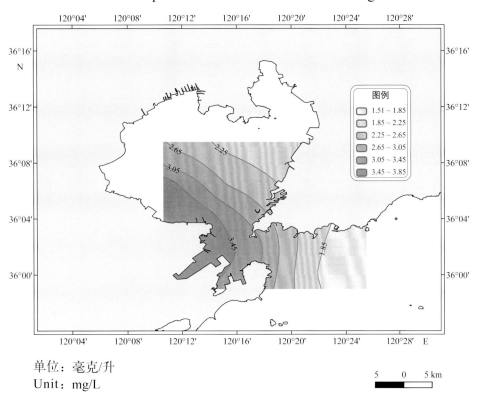

单位：毫克/升
Unit：mg/L

2013年11月悬浮体浓度分布
Distribution of suspended matter concentration in November 2013

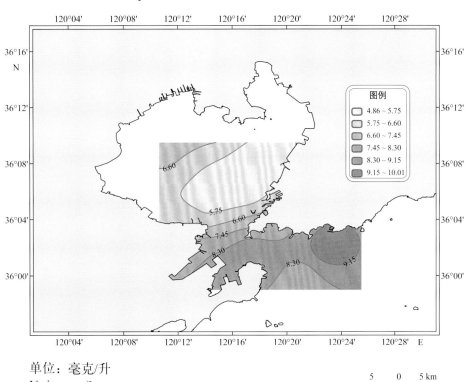

单位：毫克/升
Unit：mg/L

2014年2月悬浮体浓度分布
Distribution of suspended matter concentration in February 2014

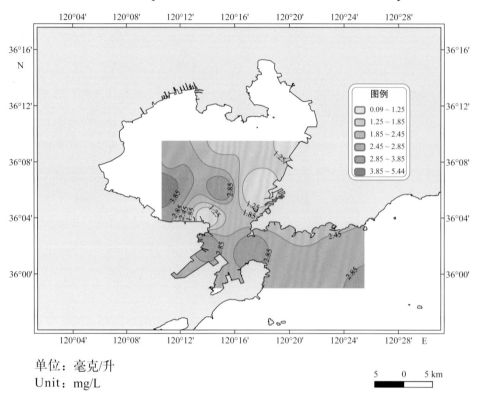

单位：毫克/升
Unit：mg/L

2014年5月悬浮体浓度分布
Distribution of suspended matter concentration in May 2014

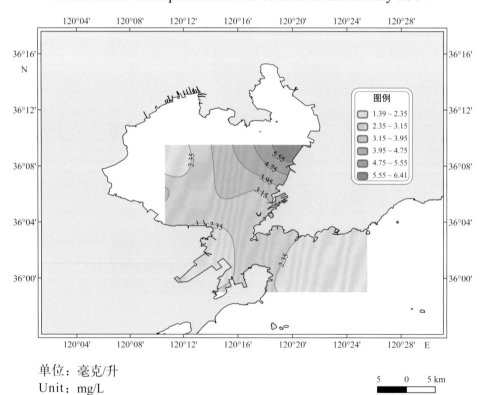

单位：毫克/升
Unit：mg/L

2014年8月悬浮体浓度分布
Distribution of suspended matter concentration in August 2014

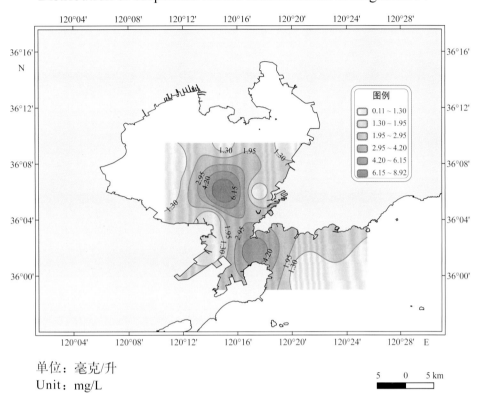

单位：毫克/升
Unit：mg/L

2014年11月悬浮体浓度分布
Distribution of suspended matter concentration in November 2014

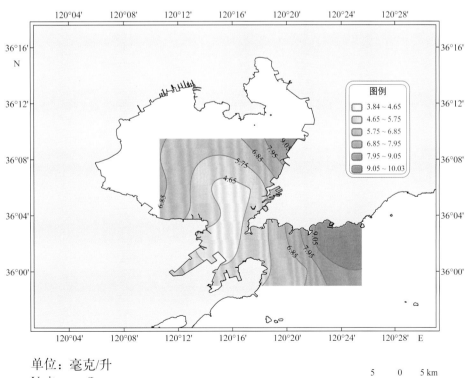

单位：毫克/升
Unit：mg/L

2015年2月悬浮体浓度分布
Distribution of suspended matter concentration in February 2015

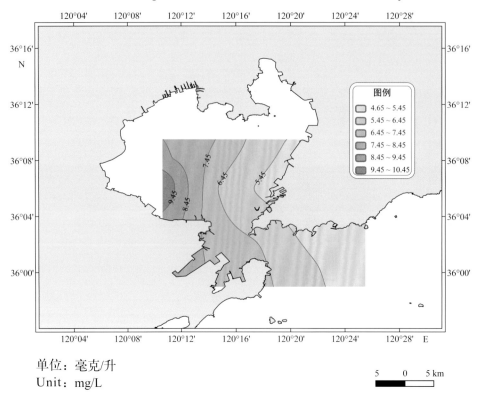

单位：毫克/升
Unit：mg/L

2015年5月悬浮体浓度分布
Distribution of suspended matter concentration in May 2015

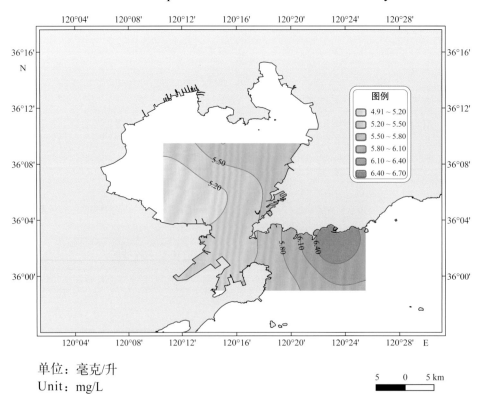

单位：毫克/升
Unit：mg/L

2015年8月悬浮体浓度分布
Distribution of suspended matter concentration in August 2015

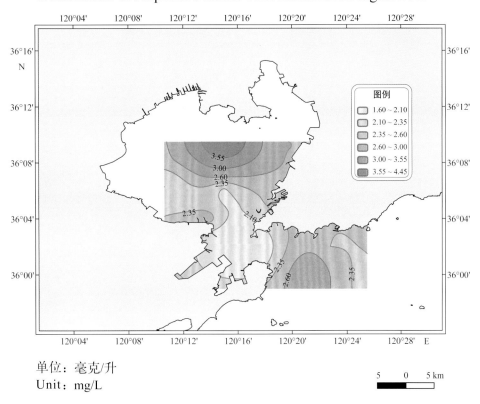

单位：毫克/升
Unit：mg/L

2015年11月悬浮体浓度分布
Distribution of suspended matter concentration in November 2015

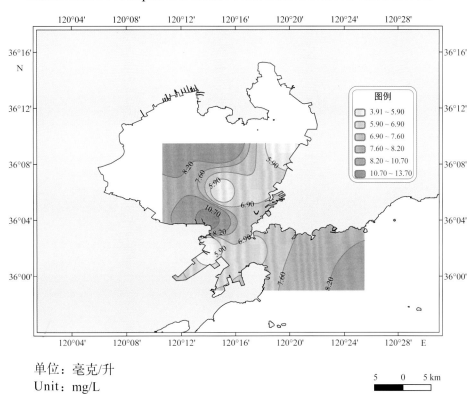

单位：毫克/升
Unit：mg/L

2. 沉积物含水率 Sediment water content

2010年2月沉积物含水率分布
Distribution of sediment water content in February 2010

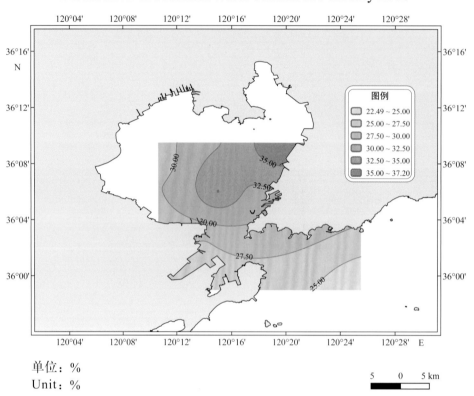

单位：%
Unit：%

2010年5月沉积物含水率分布
Distribution of sediment water content in May 2010

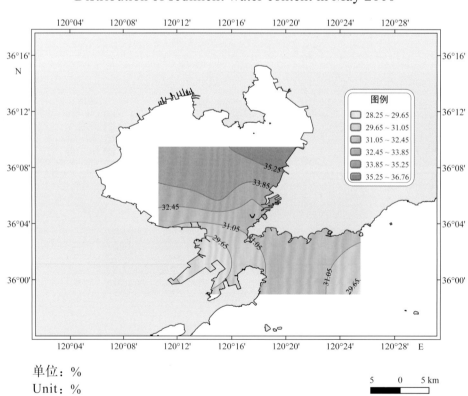

单位：%
Unit：%

2010年8月沉积物含水率分布
Distribution of sediment water content in August 2010

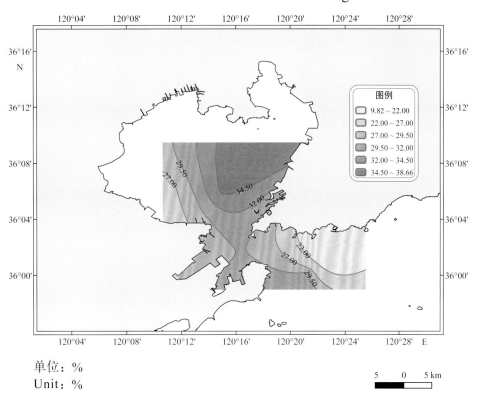

单位：%
Unit：%

2010年11月沉积物含水率分布
Distribution of sediment water content in November 2010

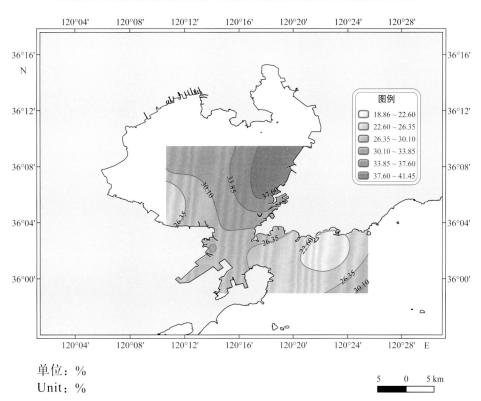

单位：%
Unit：%

2011年2月沉积物含水率分布
Distribution of sediment water content in February 2011

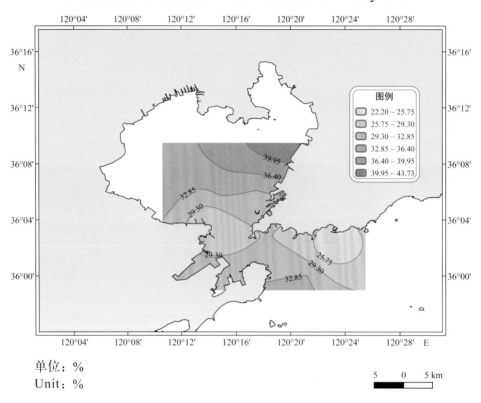

单位：%
Unit：%

2011年5月沉积物含水率分布
Distribution of sediment water content in May 2011

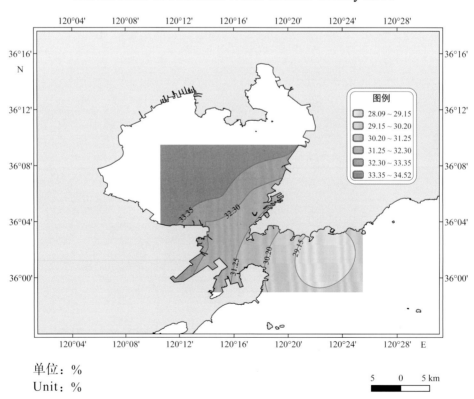

单位：%
Unit：%

2011年8月沉积物含水率分布
Distribution of sediment water content in August 2011

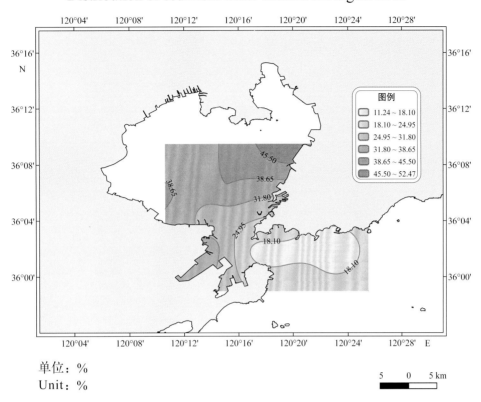

单位：%
Unit：%

2011年11月沉积物含水率分布
Distribution of sediment water content in November 2011

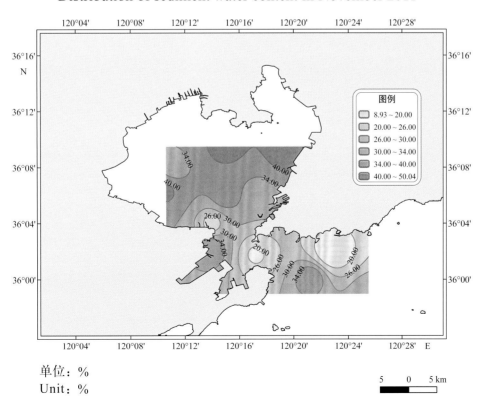

单位：%
Unit：%

2012年2月沉积物含水率分布
Distribution of sediment water content in February 2012

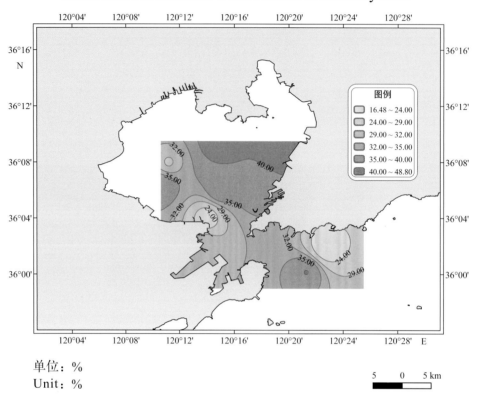

单位：%
Unit：%

2012年5月沉积物含水率分布
Distribution of sediment water content in May 2012

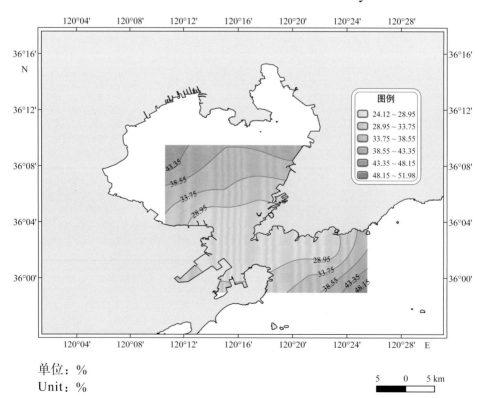

单位：%
Unit：%

2012年8月沉积物含水率分布
Distribution of sediment water content in August 2012

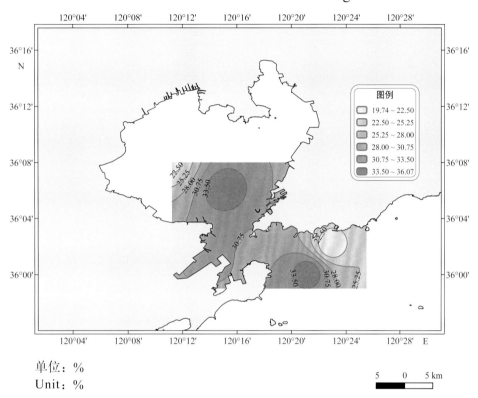

单位：%
Unit：%

2012年11月沉积物含水率分布
Distribution of sediment water content in November 2012

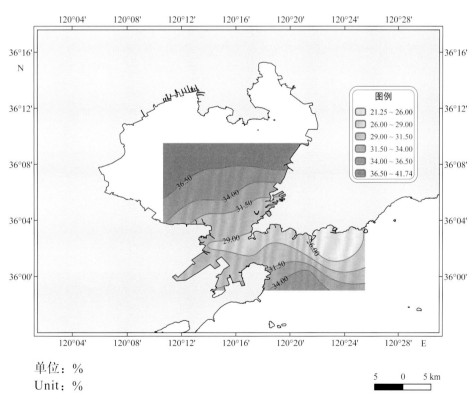

单位：%
Unit：%

2013年2月沉积物含水率分布
Distribution of sediment water content in February 2013

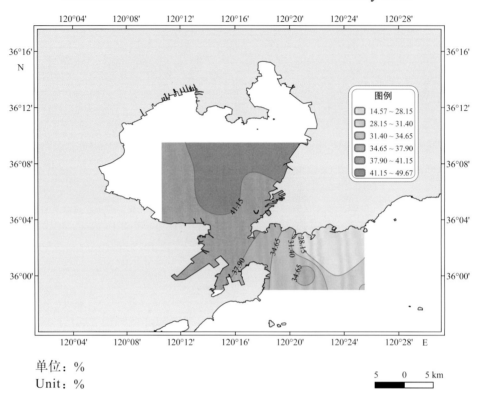

单位：%
Unit：%

2013年5月沉积物含水率分布
Distribution of sediment water content in May 2013

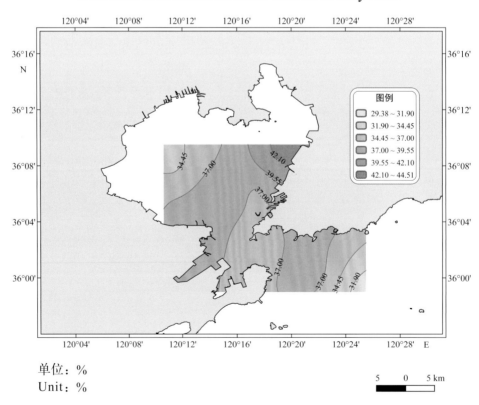

单位：%
Unit：%

2013年8月沉积物含水率分布
Distribution of sediment water content in August 2013

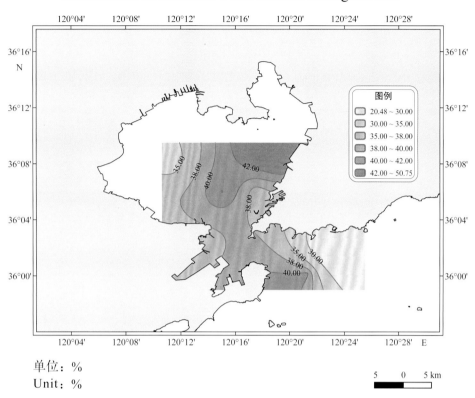

单位：%
Unit：%

2013年11月沉积物含水率分布
Distribution of sediment water content in November 2013

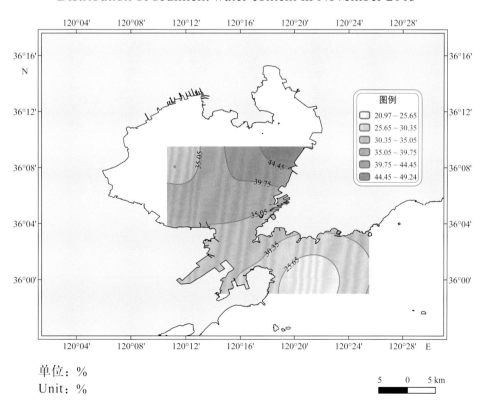

单位：%
Unit：%

2014年2月沉积物含水率分布
Distribution of sediment water content in February 2014

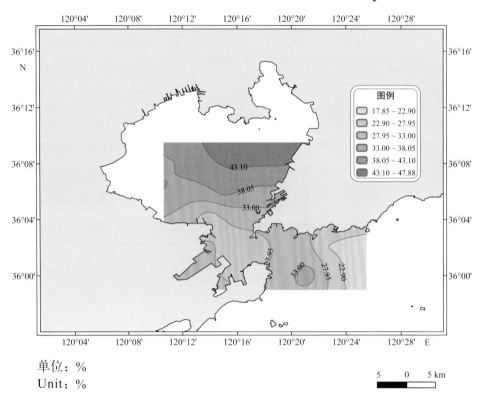

单位：%
Unit：%

2014年5月沉积物含水率分布
Distribution of sediment water content in May 2014

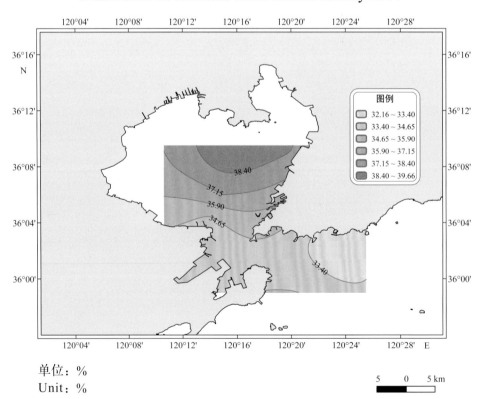

单位：%
Unit：%

2014年8月沉积物含水率分布
Distribution of sediment water content in August 2014

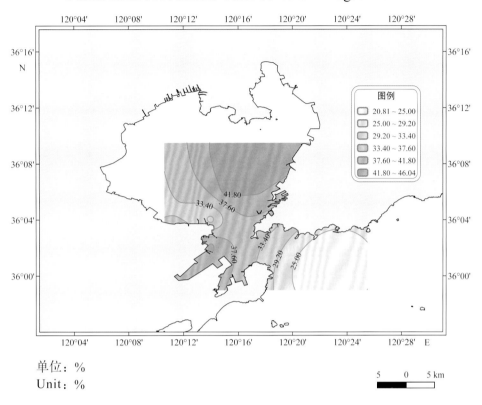

单位：%
Unit：%

2014年11月沉积物含水率分布
Distribution of sediment water content in November 2014

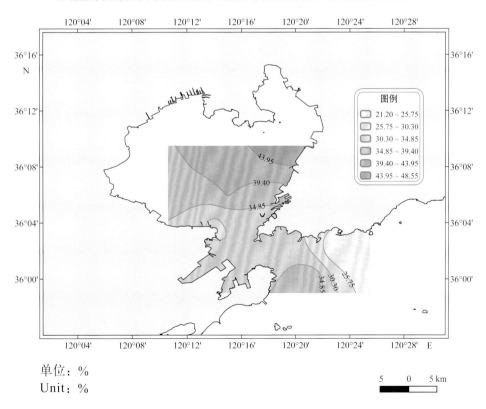

单位：%
Unit：%

2015年2月沉积物含水率分布
Distribution of sediment water content in February 2015

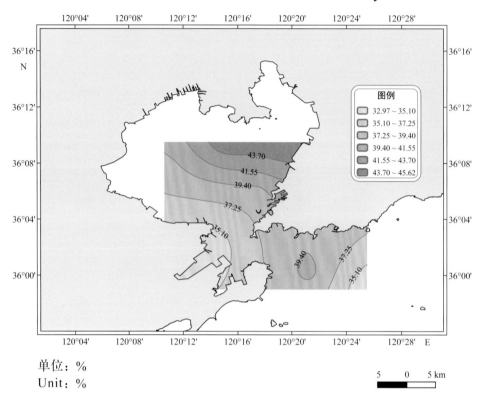

单位：%
Unit：%

2015年5月沉积物含水率分布
Distribution of sediment water content in May 2015

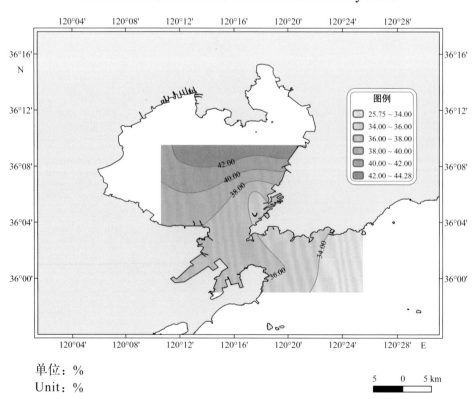

单位：%
Unit：%

2015年8月沉积物含水率分布
Distribution of sediment water content in August 2015

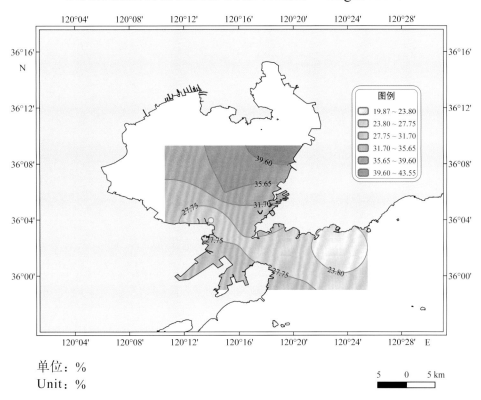

单位：%
Unit：%

2015年11月沉积物含水率分布
Distribution of sediment water content in November 2015

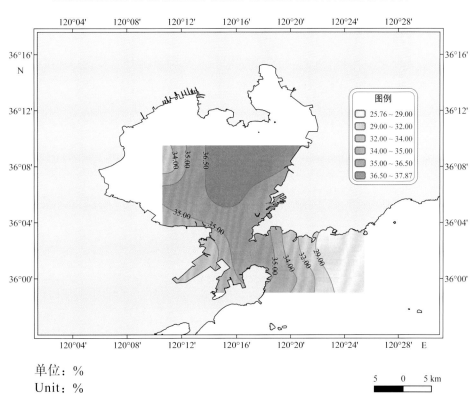

单位：%
Unit：%

3. 沉积物总磷 Total phosphorus in sediment

2010年2月沉积物总磷分布
Distribution of total phosphorus in sediment in February 2010

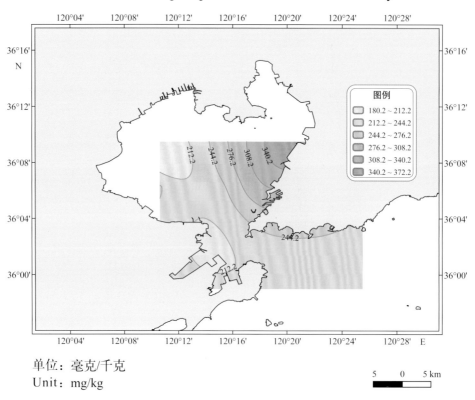

单位：毫克/千克
Unit：mg/kg

2010年5月沉积物总磷分布
Distribution of total phosphorus in sediment in May 2010

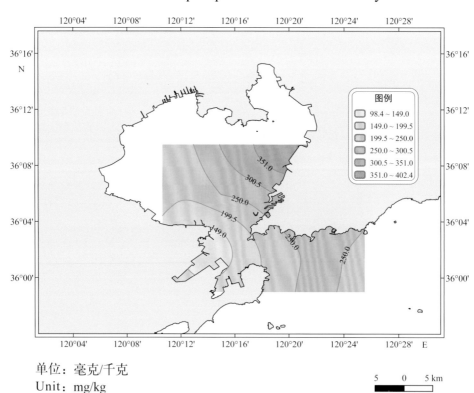

单位：毫克/千克
Unit：mg/kg

2010年8月沉积物总磷分布
Distribution of total phosphorus in sediment in August 2010

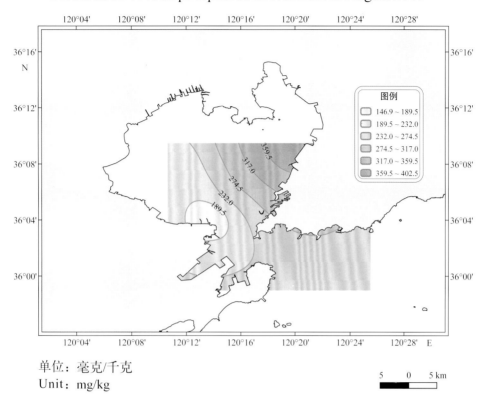

单位：毫克/千克
Unit：mg/kg

2010年11月沉积物总磷分布
Distribution of total phosphorus in sediment in November 2010

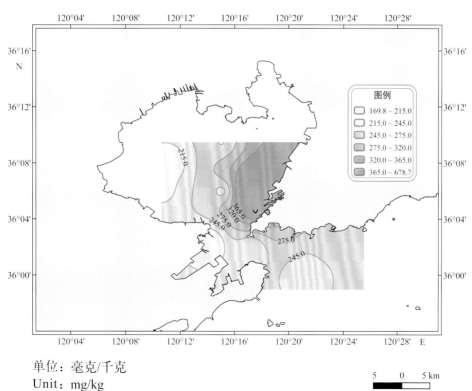

单位：毫克/千克
Unit：mg/kg

2011年2月沉积物总磷分布
Distribution of total phosphorus in sediment in February 2011

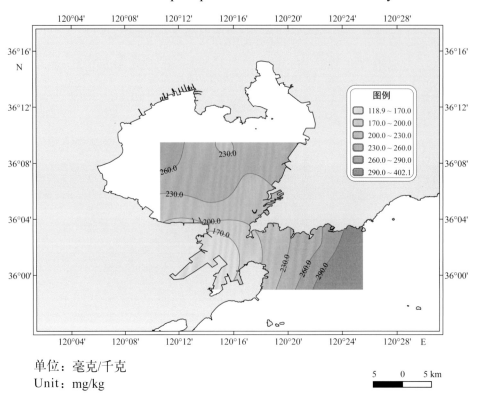

单位：毫克/千克
Unit：mg/kg

2011年5月沉积物总磷分布
Distribution of total phosphorus in sediment in May 2011

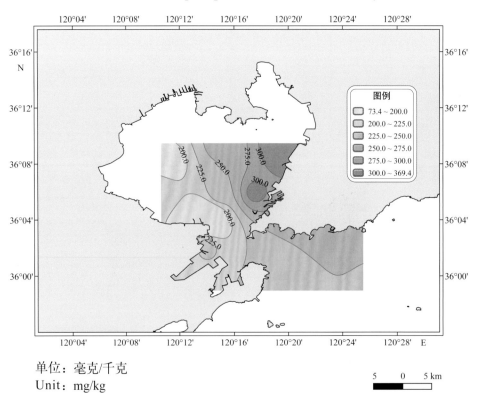

单位：毫克/千克
Unit：mg/kg

2011年8月沉积物总磷分布
Distribution of total phosphorus in sediment in August 2011

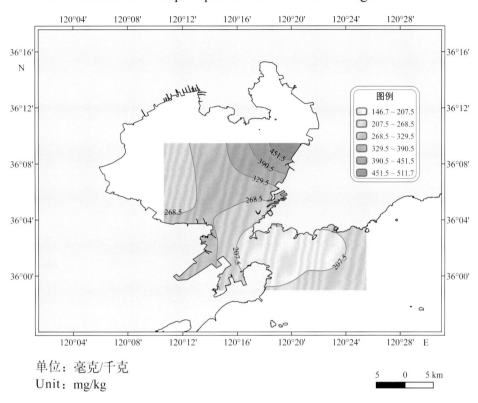

单位：毫克/千克
Unit：mg/kg

2011年11月沉积物总磷分布
Distribution of total phosphorus in sediment in November 2011

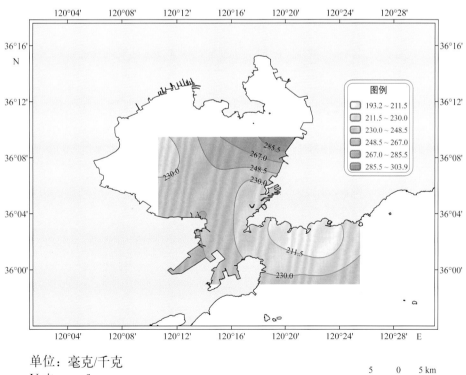

单位：毫克/千克
Unit：mg/kg

2012年2月沉积物总磷分布
Distribution of total phosphorus in sediment in February 2012

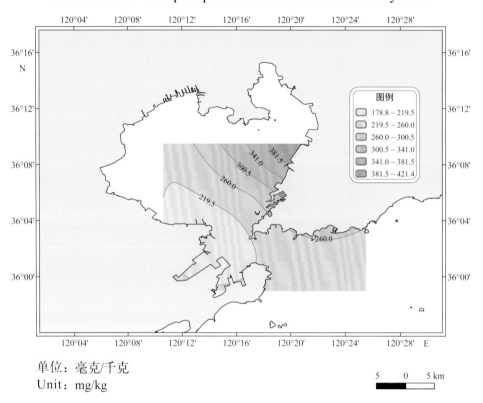

单位：毫克/千克
Unit：mg/kg

2012年5月沉积物总磷分布
Distribution of total phosphorus in sediment in May 2012

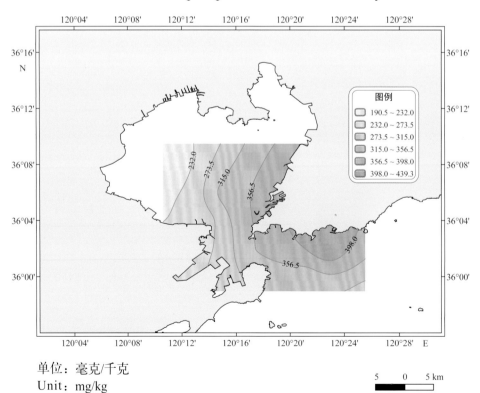

单位：毫克/千克
Unit：mg/kg

2012年8月沉积物总磷分布
Distribution of total phosphorus in sediment in August 2012

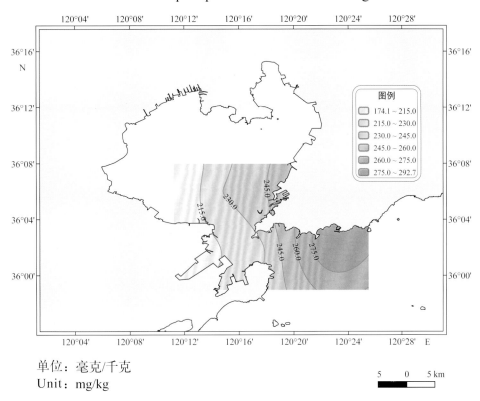

单位：毫克/千克
Unit：mg/kg

2012年11月沉积物总磷分布
Distribution of total phosphorus in sediment in November 2012

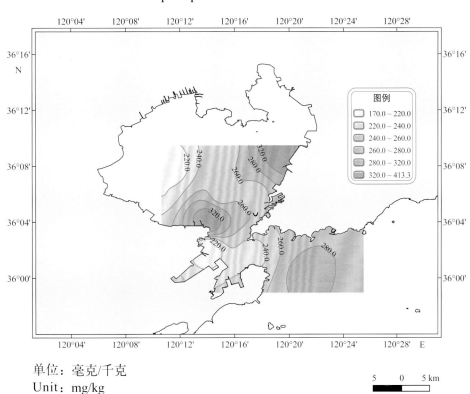

单位：毫克/千克
Unit：mg/kg

2013年2月沉积物总磷分布
Distribution of total phosphorus in sediment in February 2013

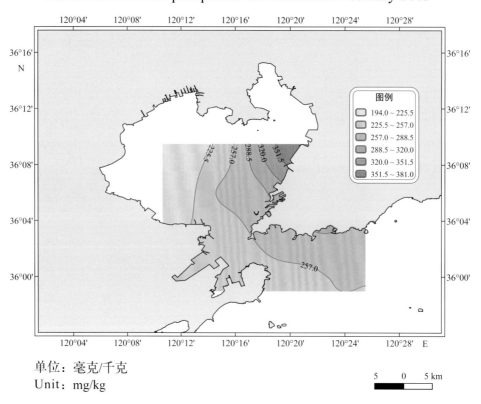

单位：毫克/千克
Unit：mg/kg

2013年5月沉积物总磷分布
Distribution of total phosphorus in sediment in May 2013

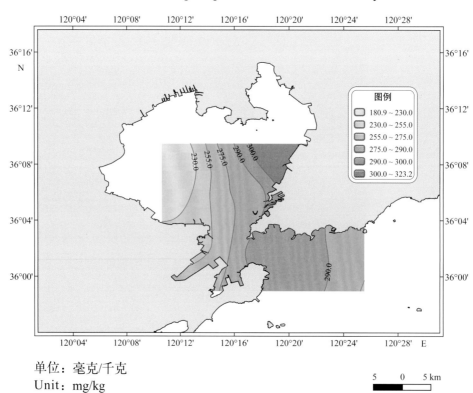

单位：毫克/千克
Unit：mg/kg

2013年8月沉积物总磷分布
Distribution of total phosphorus in sediment in August 2013

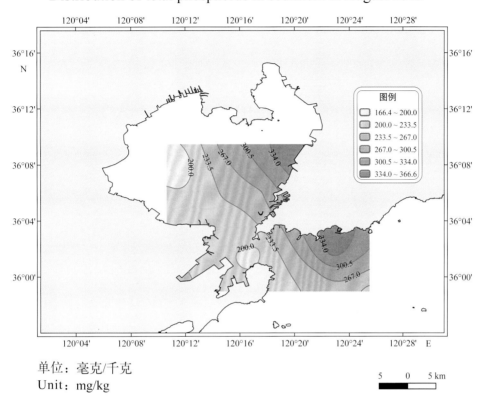

单位：毫克/千克
Unit：mg/kg

2013年11月沉积物总磷分布
Distribution of total phosphorus in sediment in November 2013

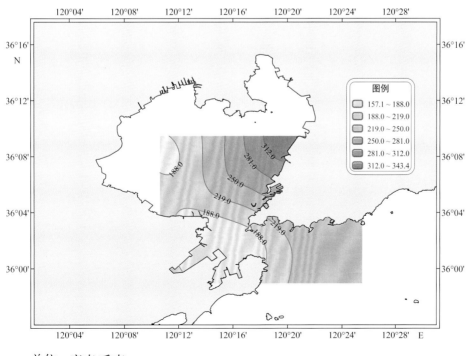

单位：毫克/千克
Unit：mg/kg

2014年2月沉积物总磷分布
Distribution of total phosphorus in sediment in February 2014

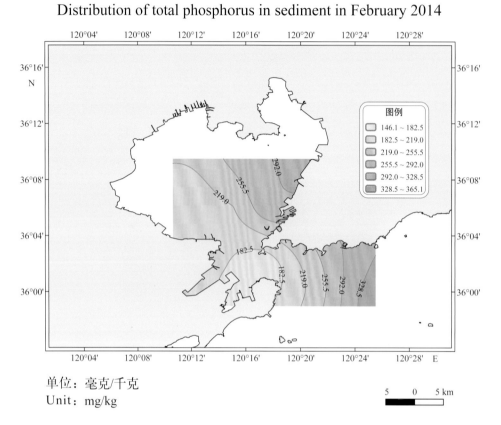

单位：毫克/千克
Unit：mg/kg

2014年5月沉积物总磷分布
Distribution of total phosphorus in sediment in May 2014

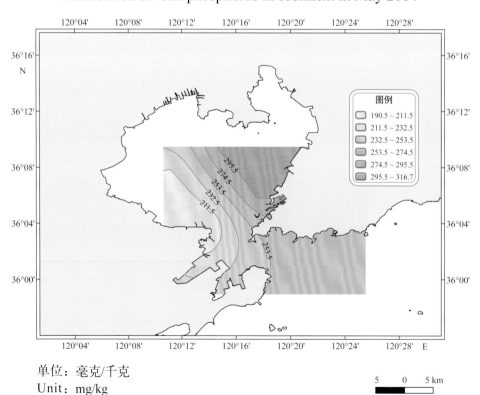

单位：毫克/千克
Unit：mg/kg

2014年8月沉积物总磷分布
Distribution of total phosphorus in sediment in August 2014

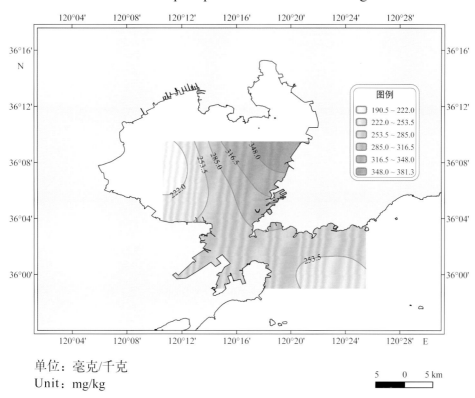

单位：毫克/千克
Unit：mg/kg

2014年11月沉积物总磷分布
Distribution of total phosphorus in sediment in November 2014

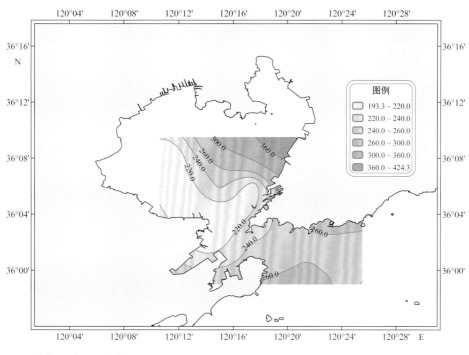

单位：毫克/千克
Unit：mg/kg

2015年2月沉积物总磷分布
Distribution of total phosphorus in sediment in February 2015

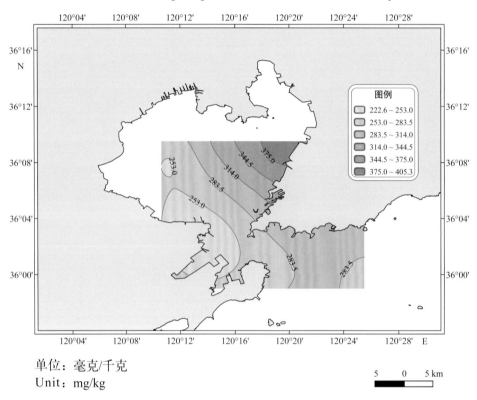

单位：毫克/千克
Unit：mg/kg

2015年5月沉积物总磷分布
Distribution of total phosphorus in sediment in May 2015

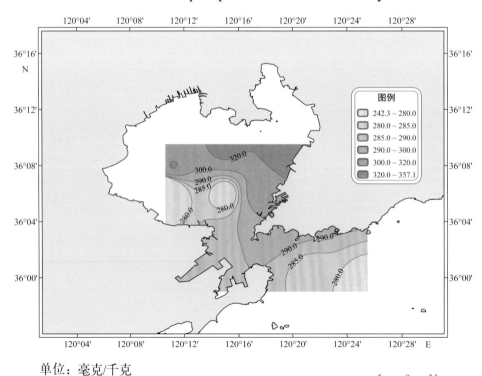

单位：毫克/千克
Unit：mg/kg

2015年8月沉积物总磷分布
Distribution of total phosphorus in sediment in August 2015

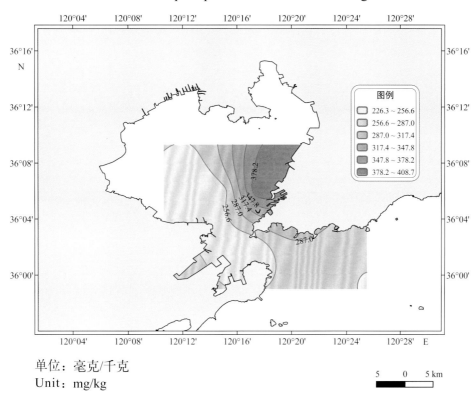

单位：毫克/千克
Unit: mg/kg

2015年11月沉积物总磷分布
Distribution of total phosphorus in sediment in November 2015

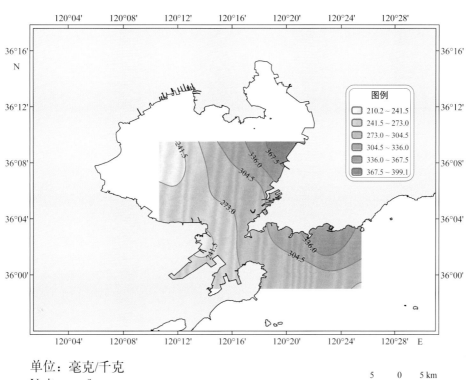

单位：毫克/千克
Unit: mg/kg

4. 沉积物总氮 Total nitrogen in sediment

2010年2月沉积物总氮分布
Distribution of total nitrogen in sediment in February 2010

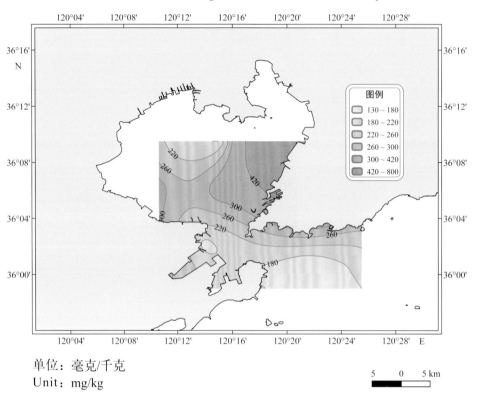

单位：毫克/千克
Unit：mg/kg

2010年5月沉积物总氮分布
Distribution of total nitrogen in sediment in May 2010

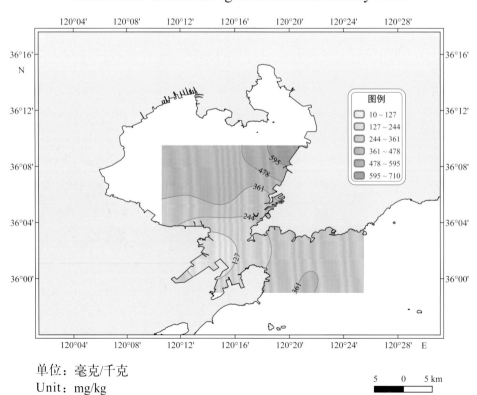

单位：毫克/千克
Unit：mg/kg

2010年8月沉积物总氮分布
Distribution of total nitrogen in sediment in August 2010

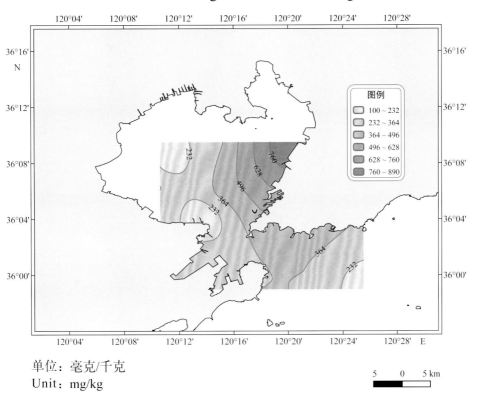

单位：毫克/千克
Unit：mg/kg

2010年11月沉积物总氮分布
Distribution of total nitrogen in sediment in November 2010

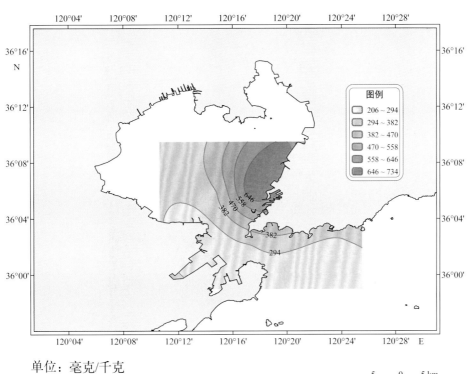

单位：毫克/千克
Unit：mg/kg

2011年2月沉积物总氮分布
Distribution of total nitrogen in sediment in February 2011

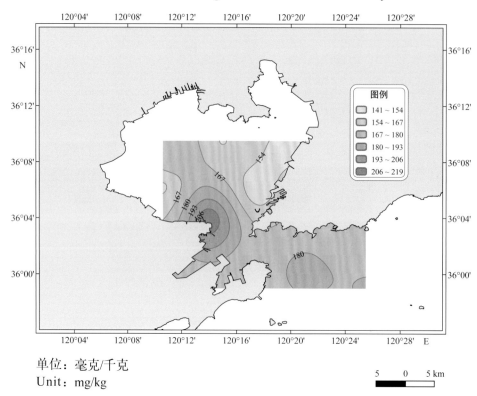

单位：毫克/千克
Unit：mg/kg

2011年5月沉积物总氮分布
Distribution of total nitrogen in sediment in May 2011

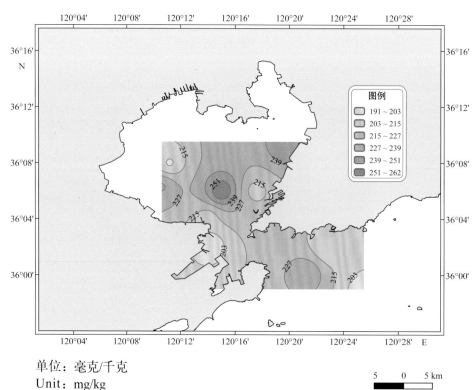

单位：毫克/千克
Unit：mg/kg

2011年8月沉积物总氮分布
Distribution of total nitrogen in sediment in August 2011

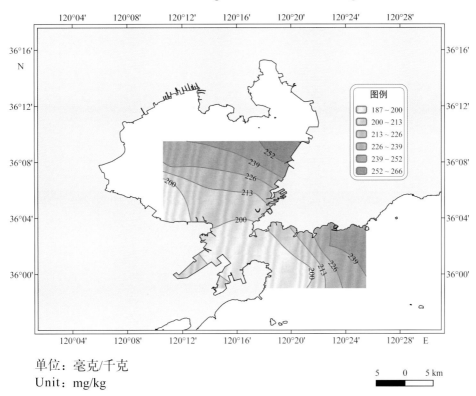

单位：毫克/千克
Unit：mg/kg

2011年11月沉积物总氮分布
Distribution of total nitrogen in sediment in November 2011

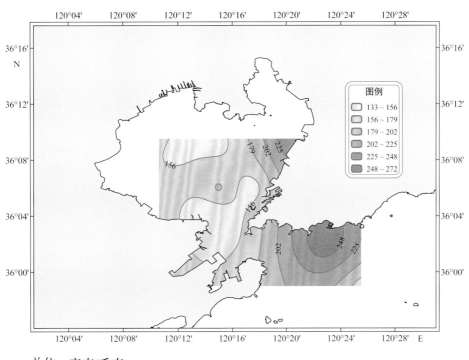

单位：毫克/千克
Unit：mg/kg

2012年2月沉积物总氮分布
Distribution of total nitrogen in sediment in February 2012

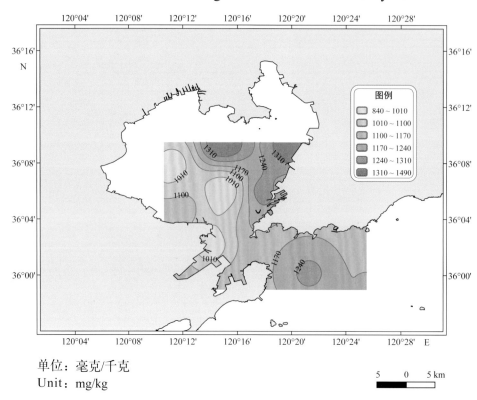

单位：毫克/千克
Unit：mg/kg

2012年5月沉积物总氮分布
Distribution of total nitrogen in sediment in May 2012

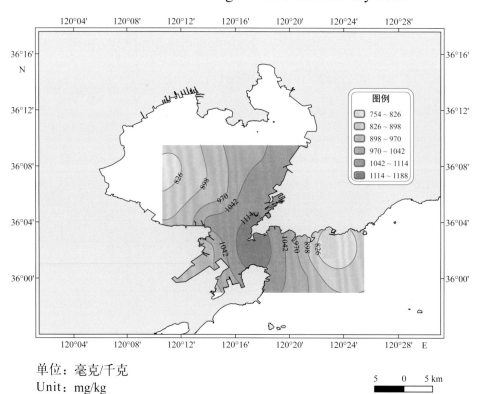

单位：毫克/千克
Unit：mg/kg

2012年8月沉积物总氮分布
Distribution of total nitrogen in sediment in August 2012

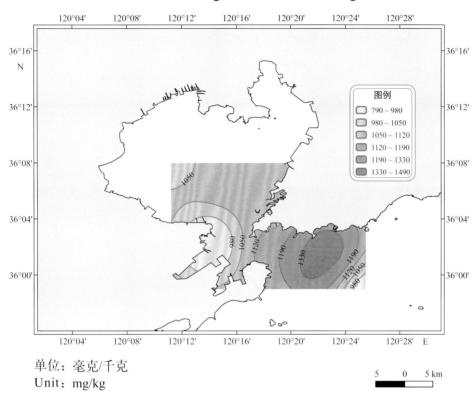

单位：毫克/千克
Unit：mg/kg

2012年11月沉积物总氮分布
Distribution of total nitrogen in sediment in November 2012

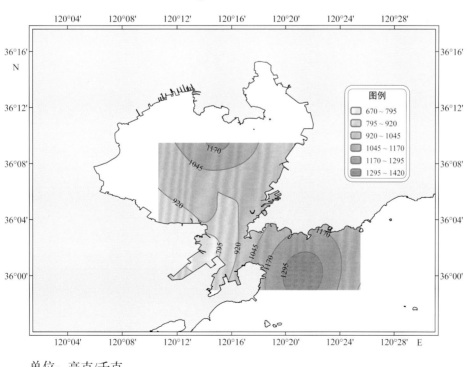

单位：毫克/千克
Unit：mg/kg

2013年2月沉积物总氮分布
Distribution of total nitrogen in sediment in February 2013

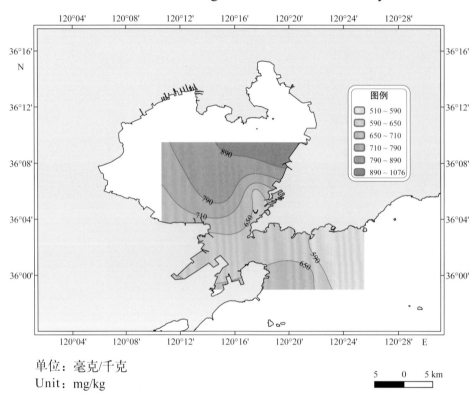

单位：毫克/千克
Unit: mg/kg

2013年5月沉积物总氮分布
Distribution of total nitrogen in sediment in May 2013

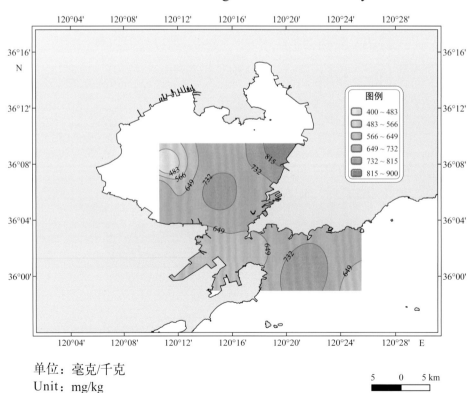

单位：毫克/千克
Unit: mg/kg

2013年8月沉积物总氮分布
Distribution of total nitrogen in sediment in August 2013

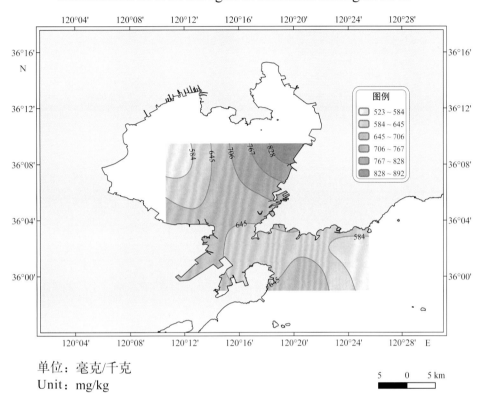

单位：毫克/千克
Unit：mg/kg

2013年11月沉积物总氮分布
Distribution of total nitrogen in sediment in November 2013

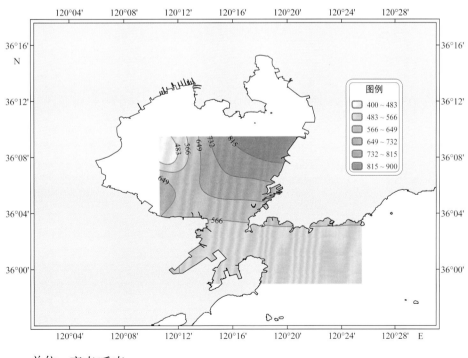

单位：毫克/千克
Unit：mg/kg

2014年2月沉积物总氮分布
Distribution of total nitrogen in sediment in February 2014

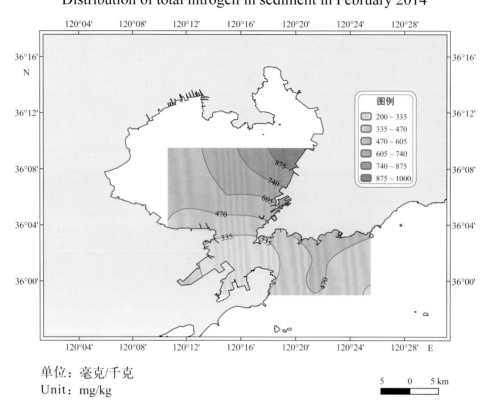

单位：毫克/千克
Unit：mg/kg

2014年5月沉积物总氮分布
Distribution of total nitrogen in sediment in May 2014

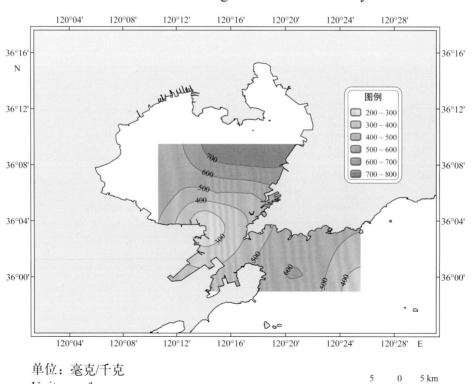

单位：毫克/千克
Unit：mg/kg

2014年8月沉积物总氮分布
Distribution of total nitrogen in sediment in August 2014

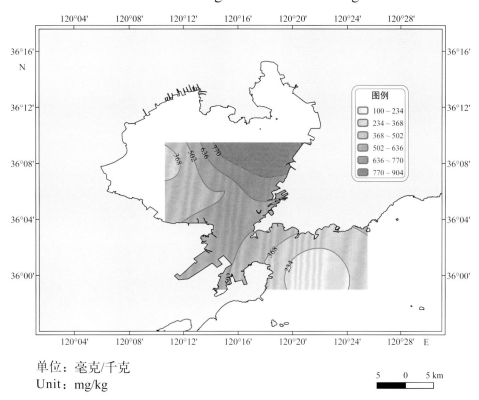

单位：毫克/千克
Unit：mg/kg

2014年11月沉积物总氮分布
Distribution of total nitrogen in sediment in November 2014

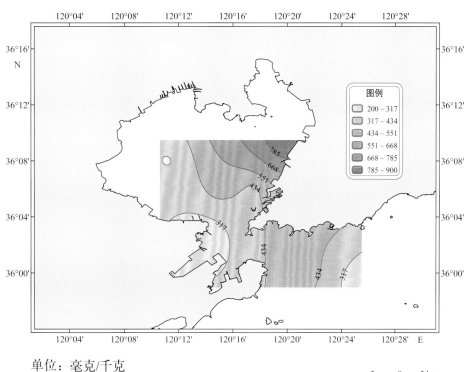

单位：毫克/千克
Unit：mg/kg

2015年2月沉积物总氮分布
Distribution of total nitrogen in sediment in February 2015

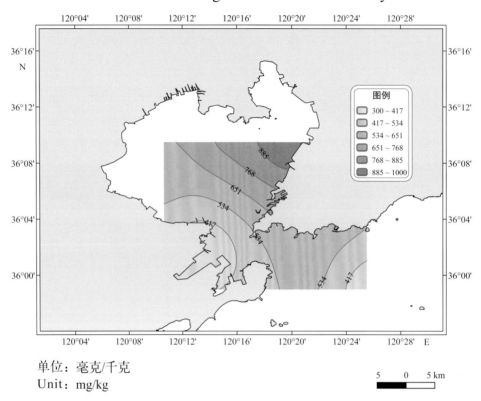

单位：毫克/千克
Unit：mg/kg

2015年5月沉积物总氮分布
Distribution of total nitrogen in sediment in May 2015

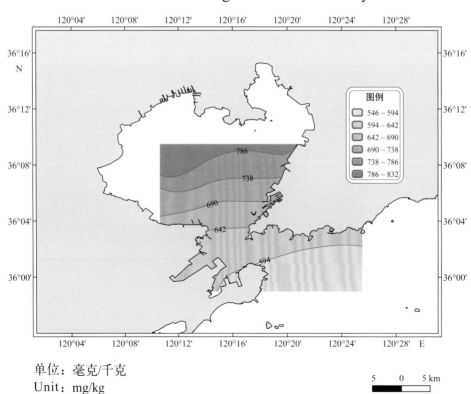

单位：毫克/千克
Unit：mg/kg

2015年8月沉积物总氮分布
Distribution of total nitrogen in sediment in August 2015

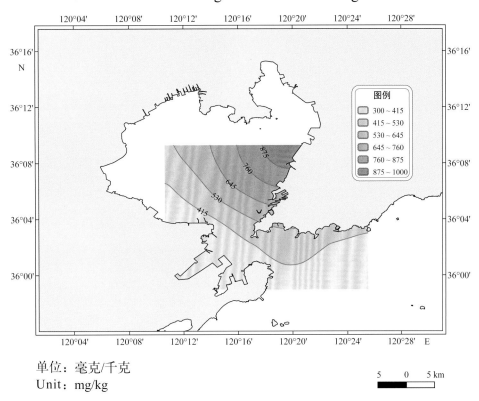

单位：毫克/千克
Unit：mg/kg

2015年11月沉积物总氮分布
Distribution of total nitrogen in sediment in November 2015

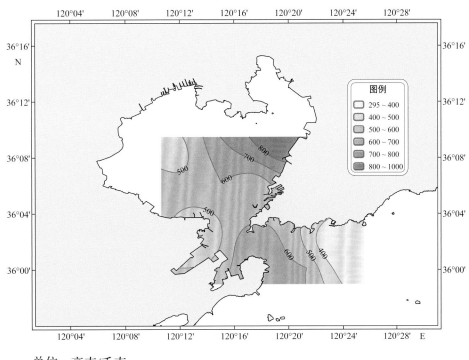

单位：毫克/千克
Unit：mg/kg

5. 沉积物砂土含量 Sand content in sediment

2010年2月沉积物砂土含量分布
Distribution of sand content in sediment in February 2010

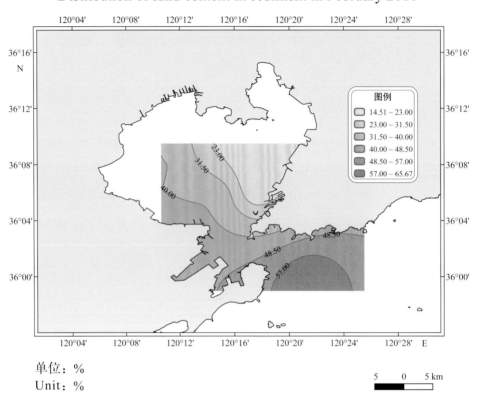

单位：%
Unit：%

2010年8月沉积物砂土含量分布
Distribution of sand content in sediment in August 2010

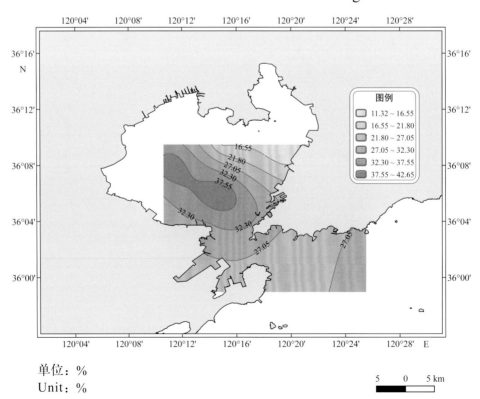

单位：%
Unit：%

2011年2月沉积物砂土含量分布
Distribution of sand content in sediment in February 2011

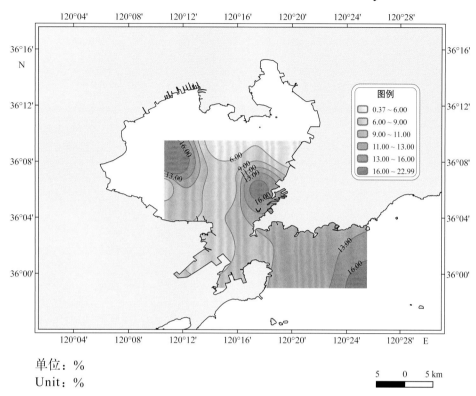

单位：%
Unit：%

2011年8月沉积物砂土含量分布
Distribution of sand content in sediment in August 2011

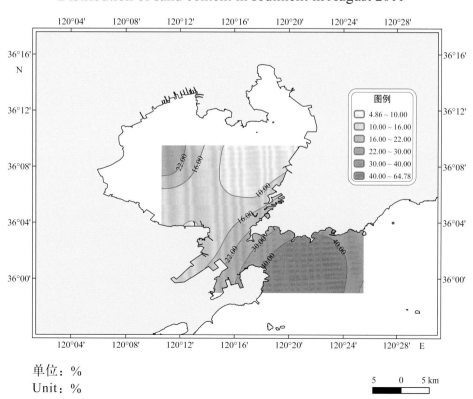

单位：%
Unit：%

2012年2月沉积物砂土含量分布
Distribution of sand content in sediment in February 2012

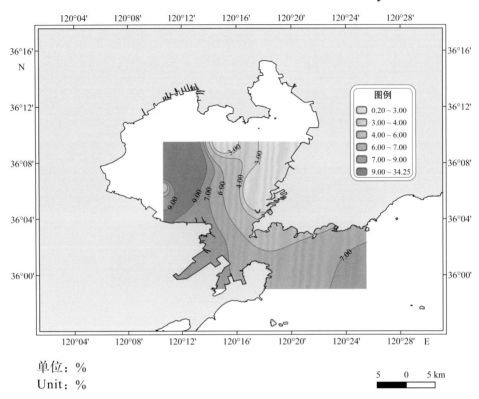

单位：%
Unit：%

2012年8月沉积物砂土含量分布
Distribution of sand content in sediment in August 2012

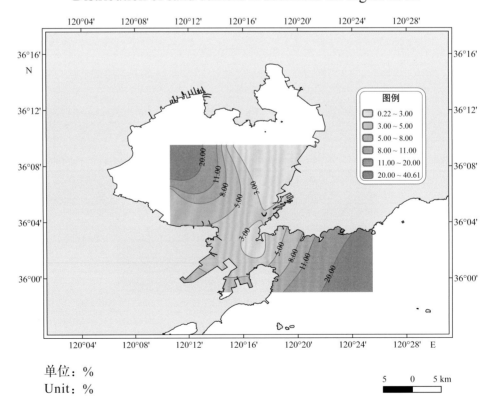

单位：%
Unit：%

2013年2月沉积物砂土含量分布
Distribution of sand content in sediment in February 2013

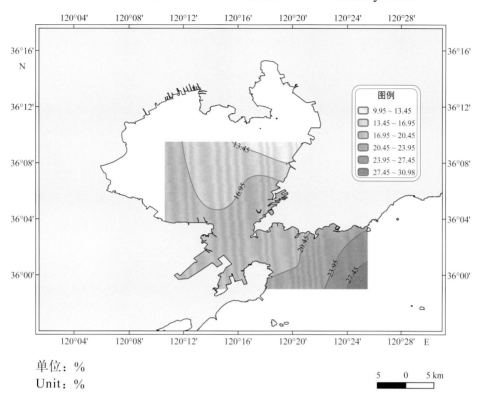

单位：%
Unit：%

2013年8月沉积物砂土含量分布
Distribution of sand content in sediment in August 2013

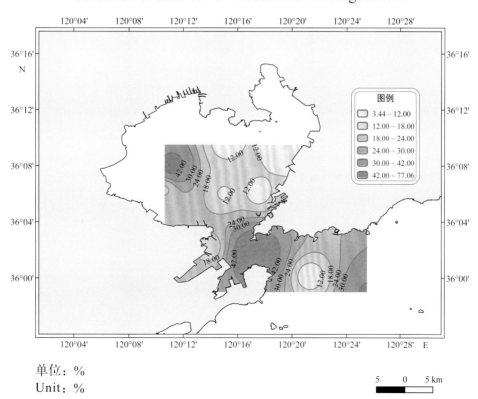

单位：%
Unit：%

2014年2月沉积物砂土含量分布
Distribution of sand content in sediment in February 2014

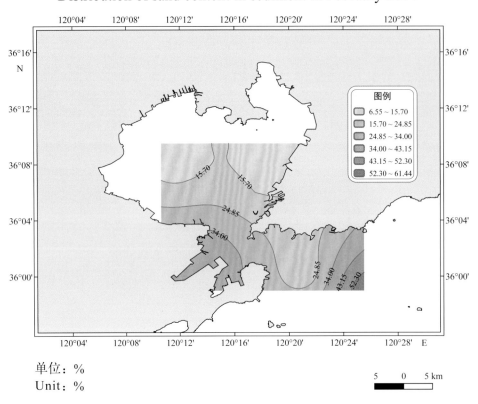

单位：%
Unit：%

2014年8月沉积物砂土含量分布
Distribution of sand content in sediment in August 2014

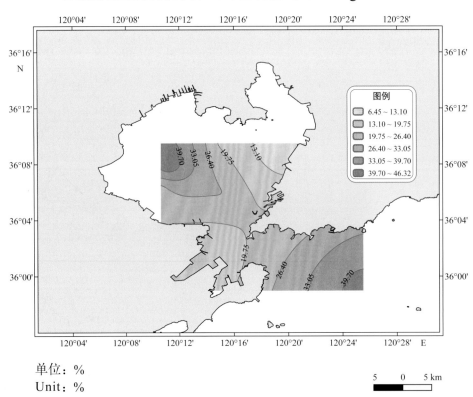

单位：%
Unit：%

2015年2月沉积物砂土含量分布
Distribution of sand content in sediment in February 2015

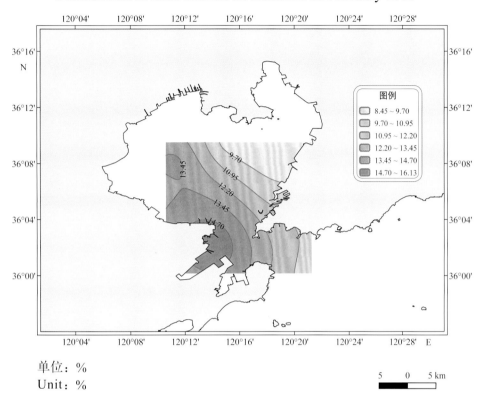

2015年8月沉积物砂土含量分布
Distribution of sand content in sediment in August 2015

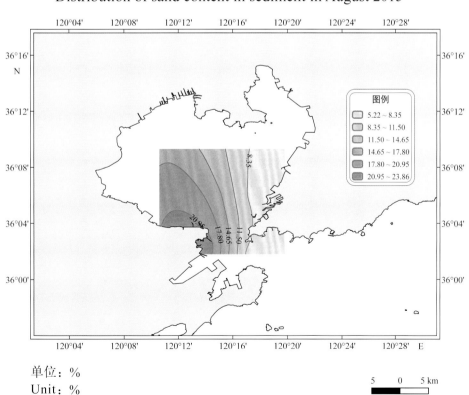

6. 沉积物粉砂含量 Silt content in sediment

2010年2月沉积物粉砂含量分布
Distribution of silt content in sediment in February 2010

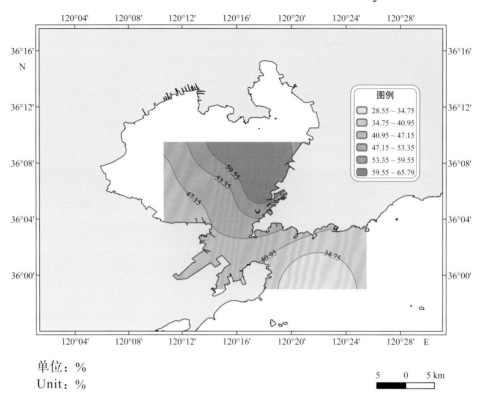

单位：%
Unit：%

2010年8月沉积物粉砂含量分布
Distribution of silt content in sediment in August 2010

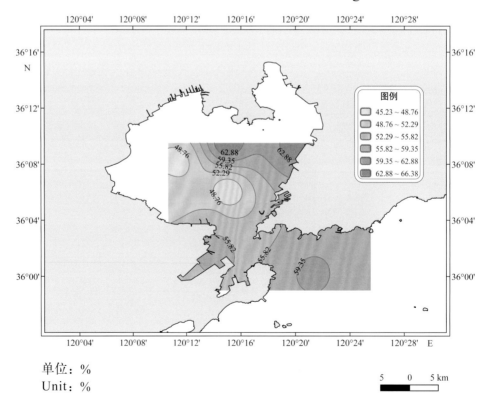

单位：%
Unit：%

2011年2月沉积物粉砂含量分布
Distribution of silt content in sediment in February 2011

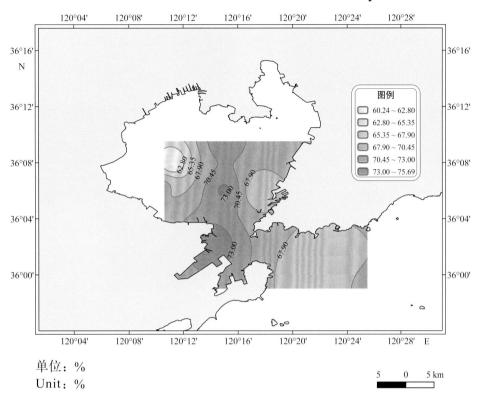

单位：%
Unit：%

2011年8月沉积物粉砂含量分布
Distribution of silt content in sediment in August 2011

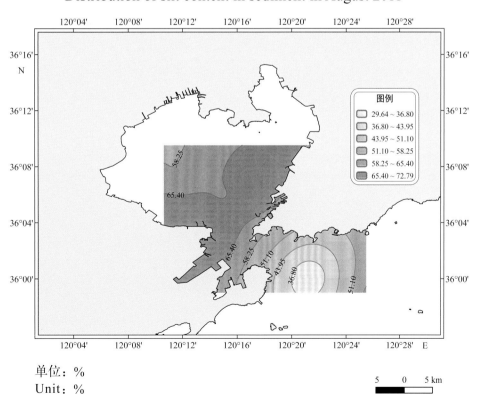

单位：%
Unit：%

2012年2月沉积物粉砂含量分布
Distribution of silt content in sediment in February 2012

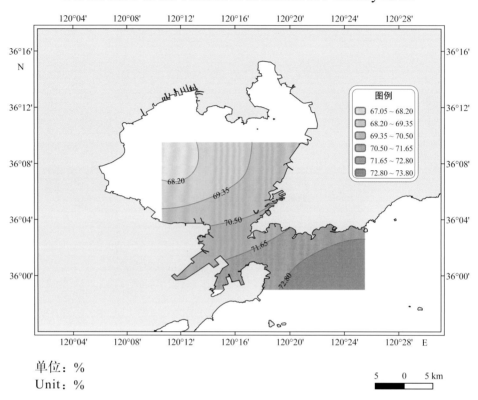

单位：%
Unit：%

2012年8月沉积物粉砂含量分布
Distribution of silt content in sediment in August 2012

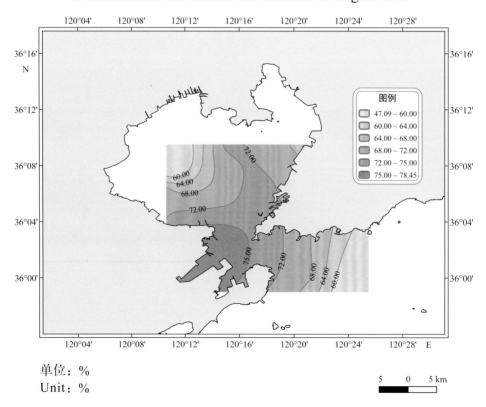

单位：%
Unit：%

2013年2月沉积物粉砂含量分布
Distribution of silt content in sediment in February 2013

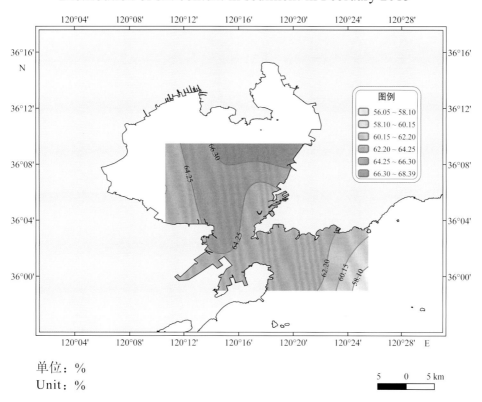

单位：%
Unit：%

2013年8月沉积物粉砂含量分布
Distribution of silt content in sediment in August 2013

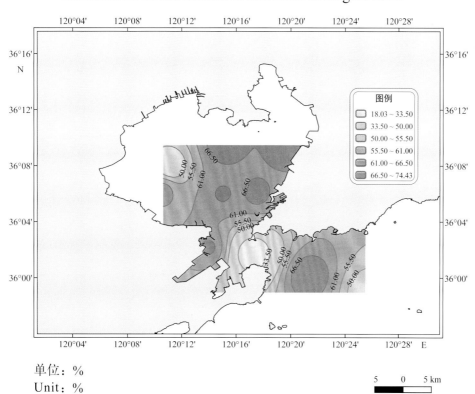

单位：%
Unit：%

2014年2月沉积物粉砂含量分布
Distribution of silt content in sediment in February 2014

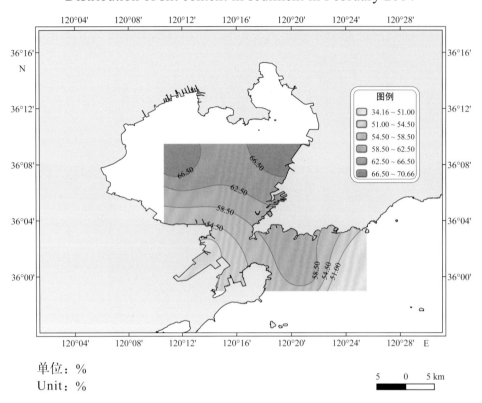

单位：%
Unit：%

2014年8月沉积物粉砂含量分布
Distribution of silt content in sediment in August 2014

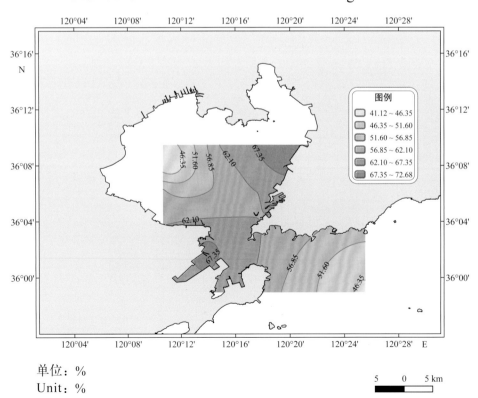

单位：%
Unit：%

2015年2月沉积物粉砂含量分布
Distribution of silt content in sediment in February 2015

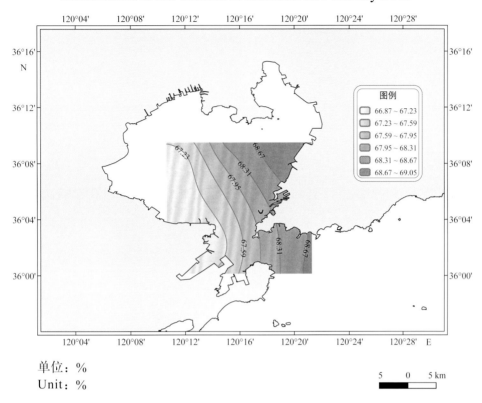

单位：%
Unit：%

2015年8月沉积物粉砂含量分布
Distribution of silt content in sediment in August 2015

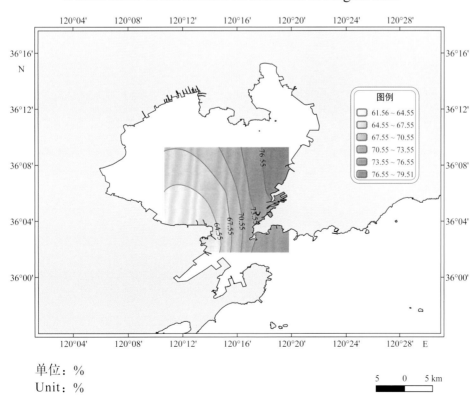

单位：%
Unit：%

7. 沉积物黏土含量 Clay content in sediment

2010年2月沉积物黏土含量分布
Distribution of clay content in sediment in February 2010

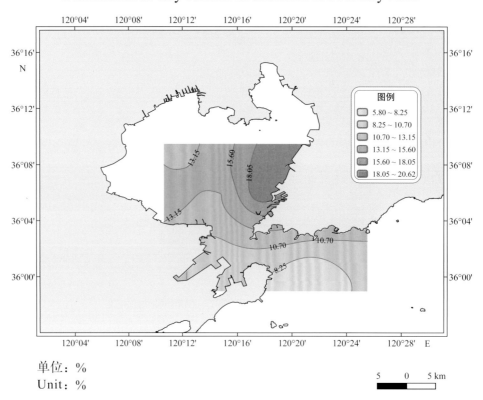

单位：%
Unit：%

2010年8月沉积物黏土含量分布
Distribution of clay content in sediment in August 2010

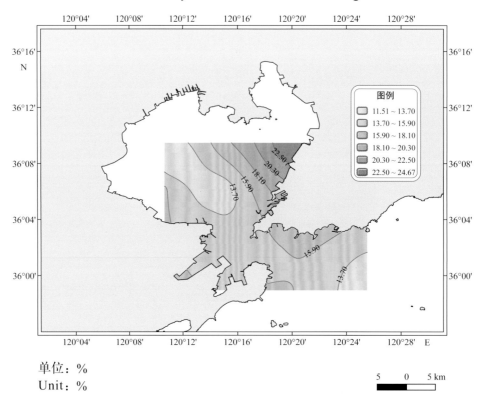

单位：%
Unit：%

2011年2月沉积物黏土含量分布
Distribution of clay content in sediment in February 2011

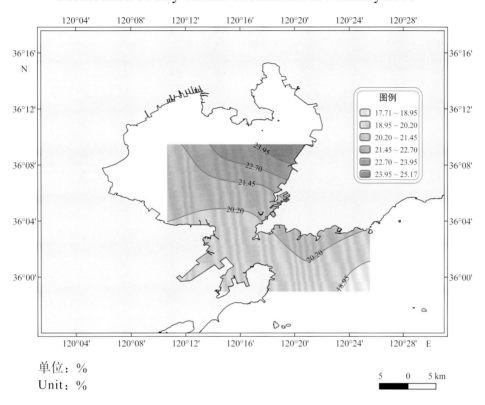

单位：%
Unit：%

2011年8月沉积物黏土含量分布
Distribution of clay content in sediment in August 2011

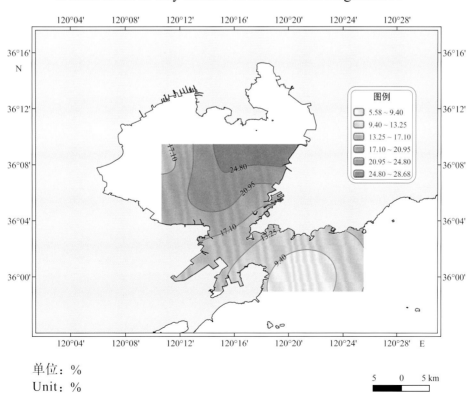

单位：%
Unit：%

2012年2月沉积物黏土含量分布
Distribution of clay content in sediment in February 2012

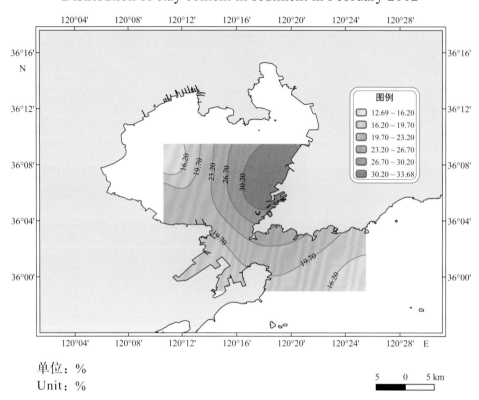

单位：%
Unit：%

2012年8月沉积物黏土含量分布
Distribution of clay content in sediment in August 2012

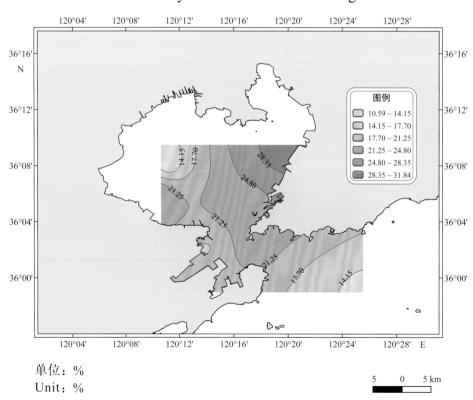

单位：%
Unit：%

2013年2月沉积物黏土含量分布
Distribution of clay content in sediment in February 2013

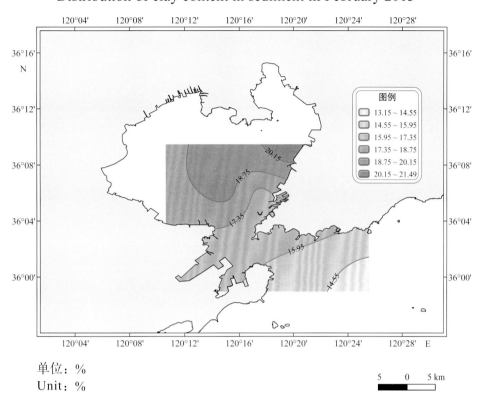

单位：%
Unit：%

2013年8月沉积物黏土含量分布
Distribution of clay content in sediment in August 2013

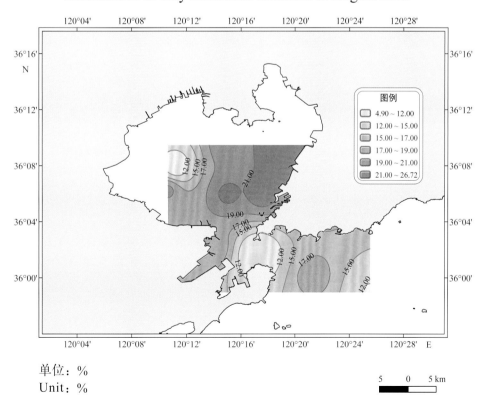

单位：%
Unit：%

2014年2月沉积物黏土含量分布
Distribution of clay content in sediment in February 2014

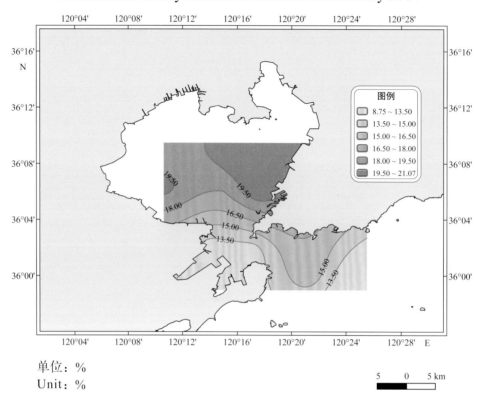

2014年8月沉积物黏土含量分布
Distribution of clay content in sediment in August 2014

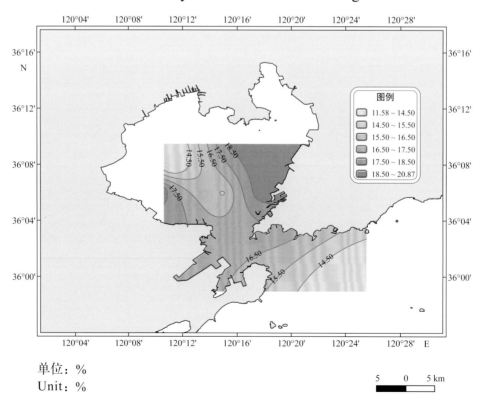

2015年2月沉积物黏土含量分布
Distribution of clay content in sediment in February 2015

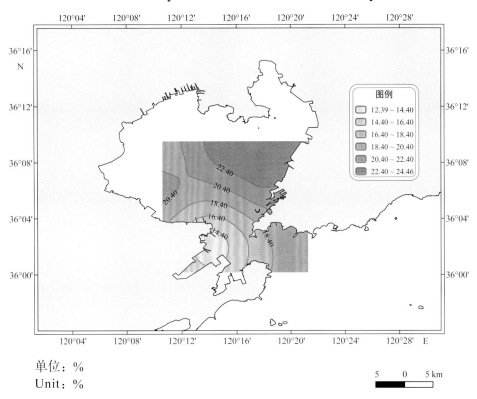

单位：%
Unit：%

2015年8月沉积物黏土含量分布
Distribution of clay content in sediment in August 2015

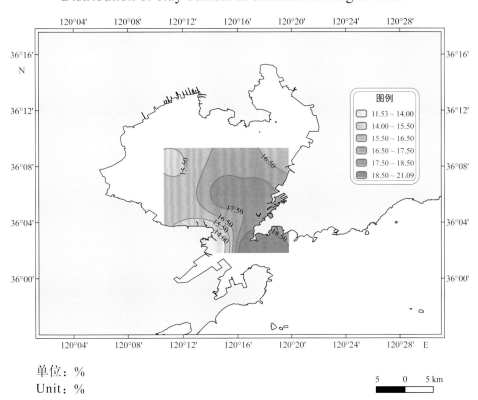

单位：%
Unit：%

8. 沉积物有机质含量 Organic matter content in sediment

2010年2月沉积物有机质含量分布
Distribution of organic matter content in sediment in February 2010

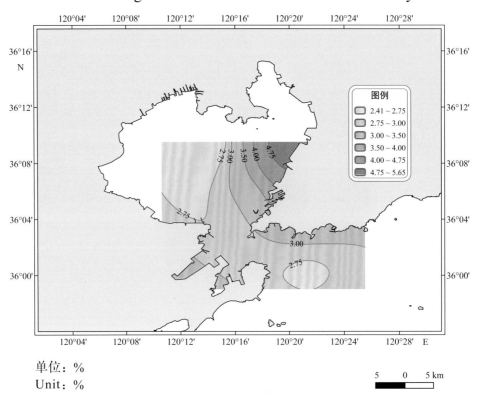

单位：%
Unit：%

2010年5月沉积物有机质含量分布
Distribution of organic matter content in sediment in May 2010

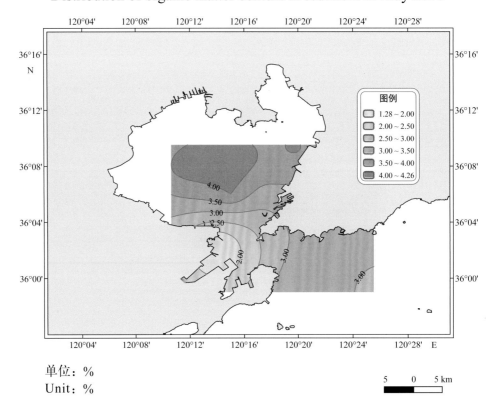

单位：%
Unit：%

2010年8月沉积物有机质含量分布
Distribution of organic matter content in sediment in August 2010

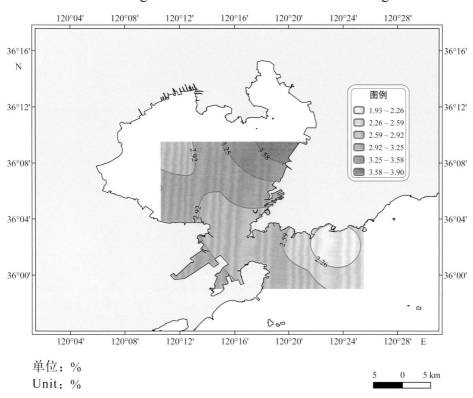

单位：%
Unit：%

2010年11月沉积物有机质含量分布
Distribution of organic matter content in sediment in November 2010

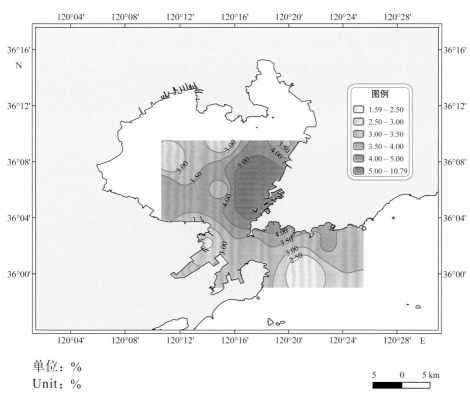

单位：%
Unit：%

2011年2月沉积物有机质含量分布
Distribution of organic matter content in sediment in February 2011

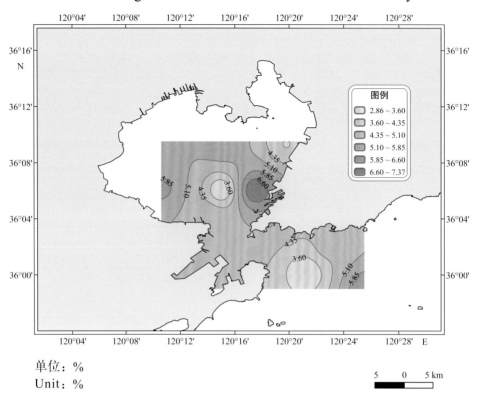

单位：%
Unit：%

2011年5月沉积物有机质含量分布
Distribution of organic matter content in sediment in May 2011

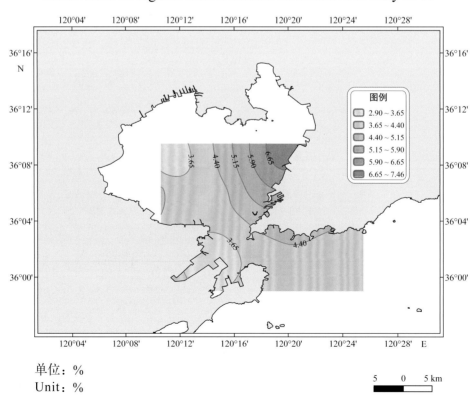

单位：%
Unit：%

2011年8月沉积物有机质含量分布
Distribution of organic matter content in sediment in August 2011

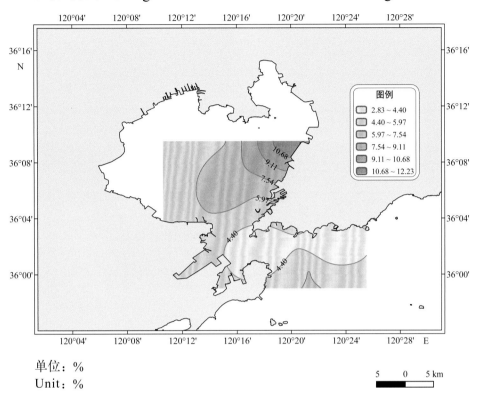

单位：%
Unit：%

2011年11月沉积物有机质含量分布
Distribution of organic matter content in sediment in November 2011

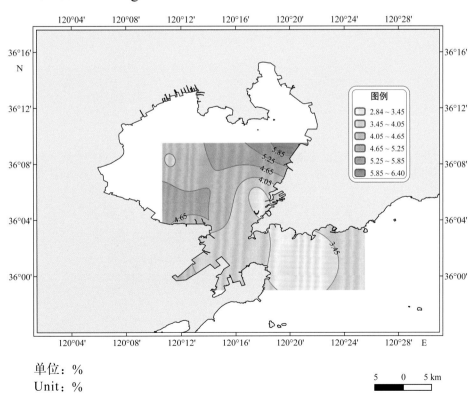

单位：%
Unit：%

2012年2月沉积物有机质含量分布
Distribution of organic matter content in sediment in February 2012

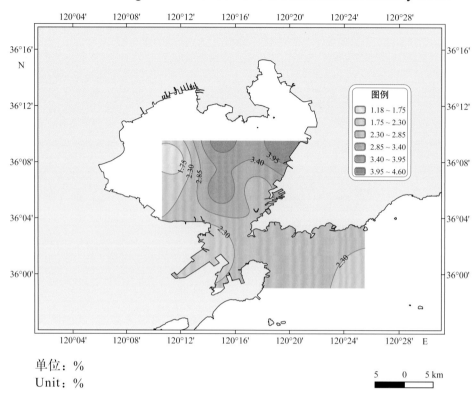

单位：%
Unit: %

2012年5月沉积物有机质含量分布
Distribution of organic matter content in sediment in May 2012

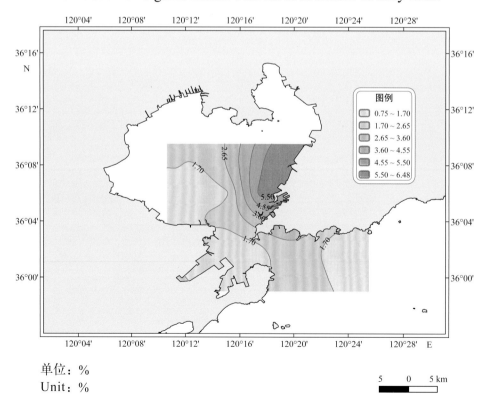

单位：%
Unit: %

2012年8月沉积物有机质含量分布
Distribution of organic matter content in sediment in August 2012

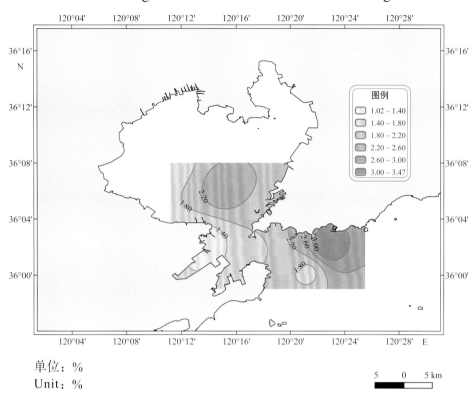

单位：%
Unit：%

2012年11月沉积物有机质含量分布
Distribution of organic matter content in sediment in November 2012

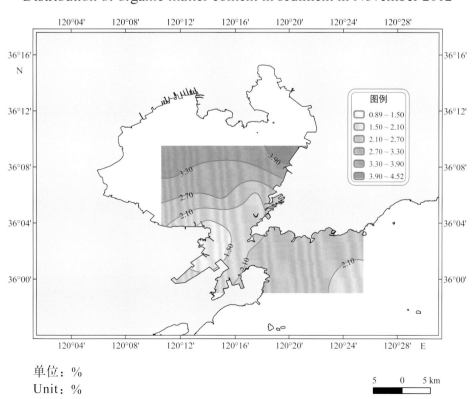

单位：%
Unit：%

2013年2月沉积物有机质含量分布
Distribution of organic matter content in sediment in February 2013

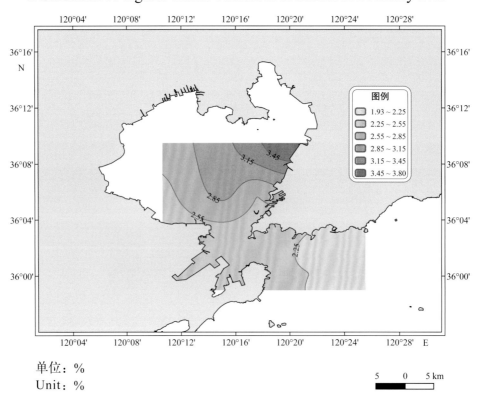

单位：%
Unit：%

2013年5月沉积物有机质含量分布
Distribution of organic matter content in sediment in May 2013

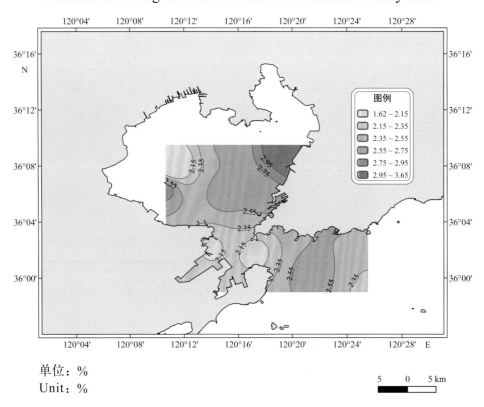

单位：%
Unit：%

2013年8月沉积物有机质含量分布
Distribution of organic matter content in sediment in August 2013

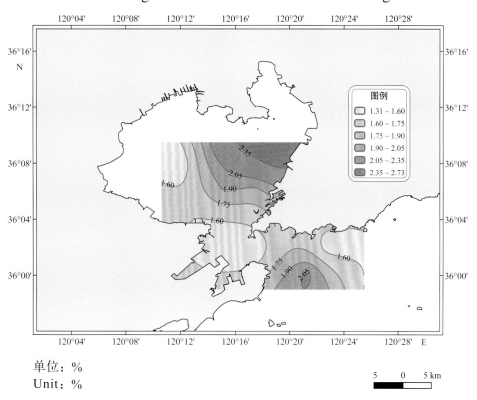

单位：%
Unit：%

2013年11月沉积物有机质含量分布
Distribution of organic matter content in sediment in November 2013

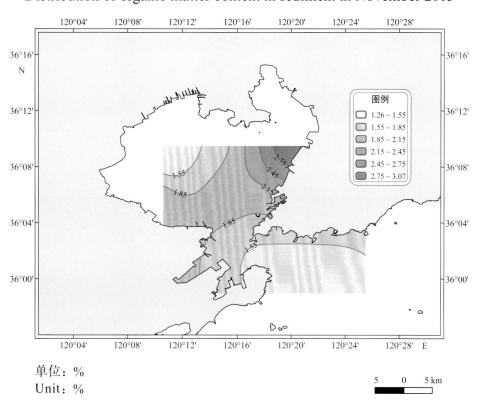

单位：%
Unit：%

2014年2月沉积物有机质含量分布
Distribution of organic matter content in sediment in February 2014

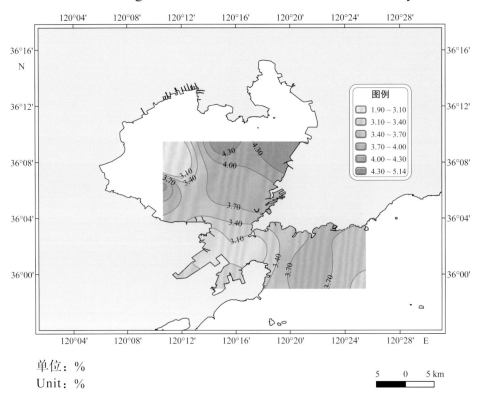

单位：%
Unit：%

2014年5月沉积物有机质含量分布
Distribution of organic matter content in sediment in May 2014

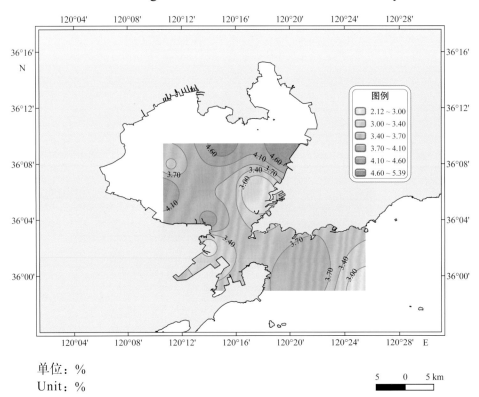

单位：%
Unit：%

2014年8月沉积物有机质含量分布
Distribution of organic matter content in sediment in August 2014

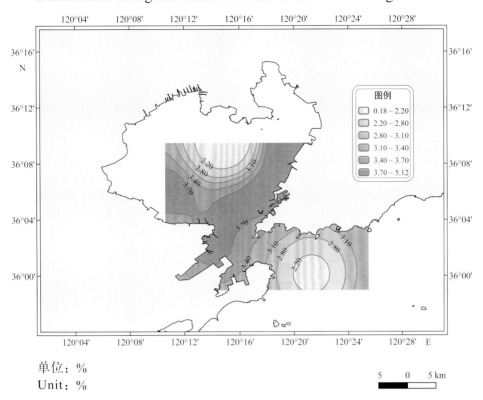

单位：%
Unit：%

2014年11月沉积物有机质含量分布
Distribution of organic matter content in sediment in November 2014

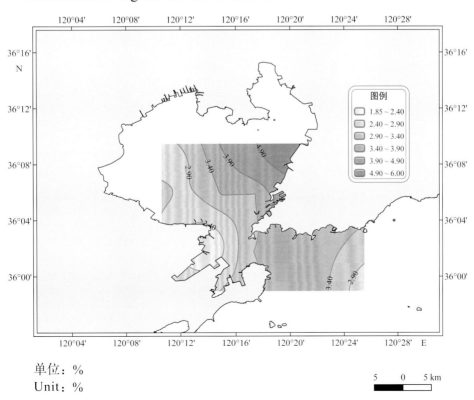

单位：%
Unit：%

2015年2月沉积物有机质含量分布
Distribution of organic matter content in sediment in February 2015

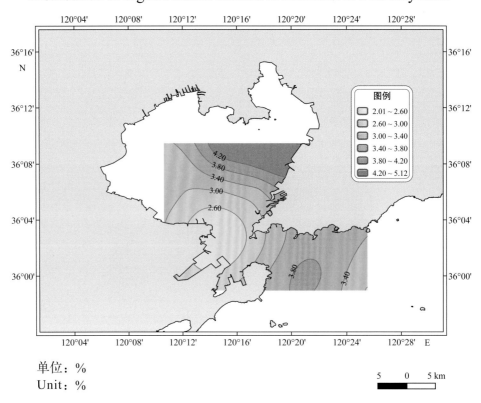

单位：%
Unit：%

2015年5月沉积物有机质含量分布
Distribution of organic matter content in sediment in May 2015

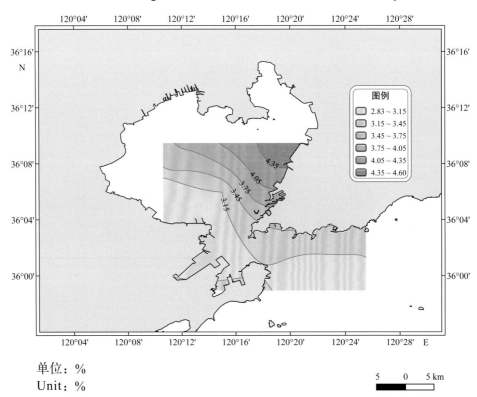

单位：%
Unit：%

2015年8月沉积物有机质含量分布
Distribution of organic matter content in sediment in August 2015

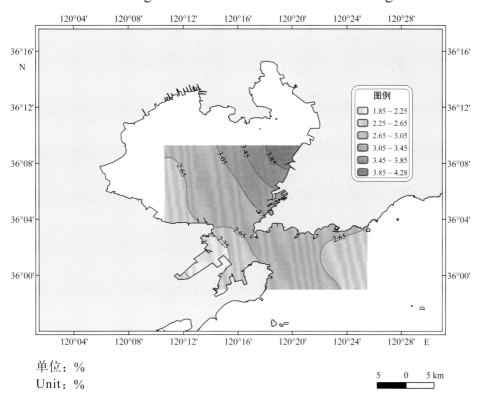

单位：%
Unit：%

2015年11月沉积物有机质含量分布
Distribution of organic matter content in sediment in November 2015

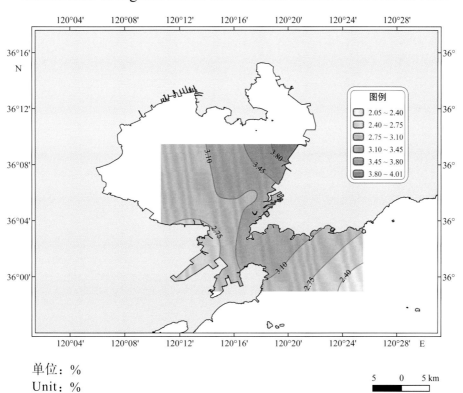

单位：%
Unit：%

四、生物海洋

BIOLOGICAL OCEANOGRAPHY

1. 叶绿素 a 浓度 Chlorophyll a concentration

2010年2月叶绿素 a 浓度分布
Distribution of chlorophyll a concentration in February 2010

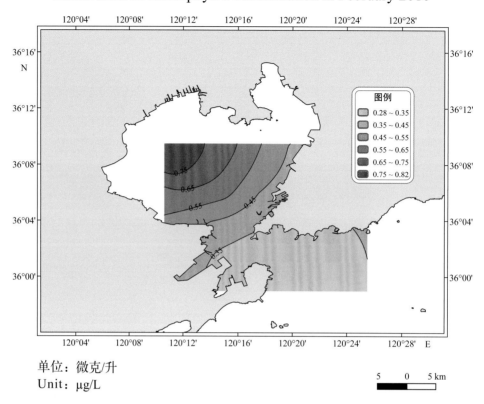

单位：微克/升
Unit：μg/L

2010年5月叶绿素 a 浓度分布
Distribution of chlorophyll a concentration in May 2010

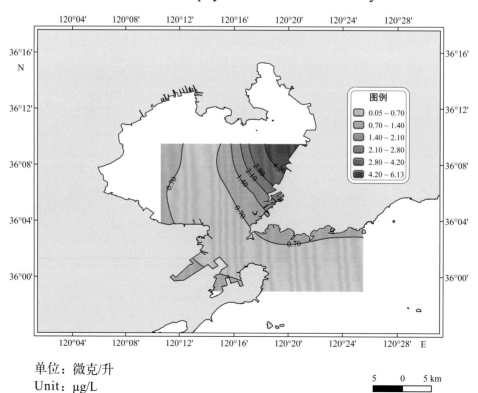

单位：微克/升
Unit：μg/L

2010年8月叶绿素 a 浓度分布
Distribution of chlorophyll a concentration in August 2010

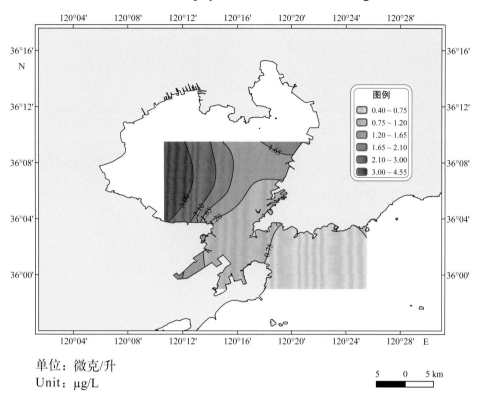

2010年11月叶绿素 a 浓度分布
Distribution of chlorophyll a concentration in November 2010

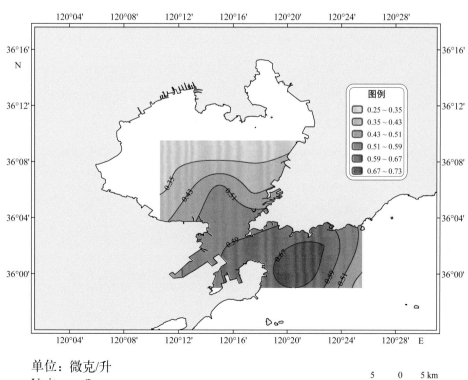

2011年2月叶绿素 a 浓度分布
Distribution of chlorophyll a concentration in February 2011

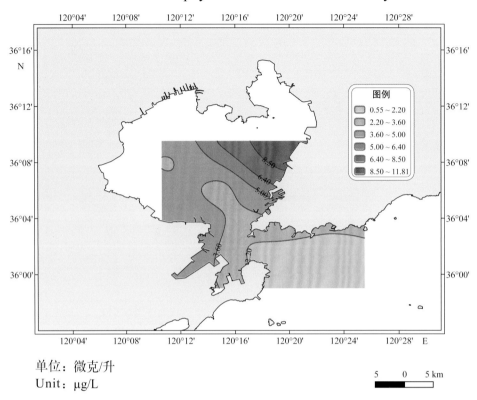

单位：微克/升
Unit：μg/L

2011年5月叶绿素 a 浓度分布
Distribution of chlorophyll a concentration in May 2011

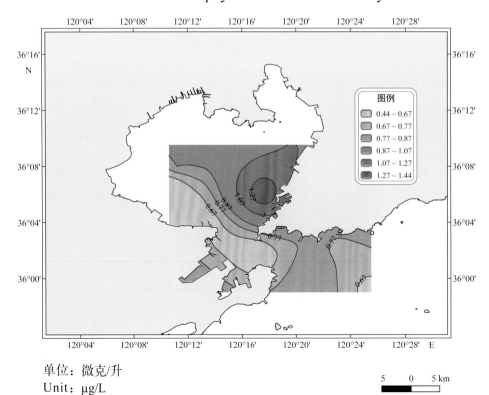

单位：微克/升
Unit：μg/L

2011年8月叶绿素a浓度分布
Distribution of chlorophyll a concentration in August 2011

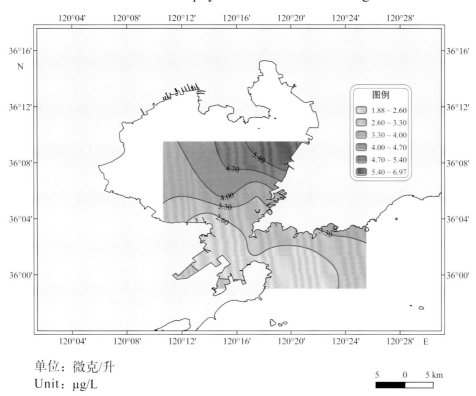

单位：微克/升
Unit：μg/L

2011年11月叶绿素a浓度分布
Distribution of chlorophyll a concentration in November 2011

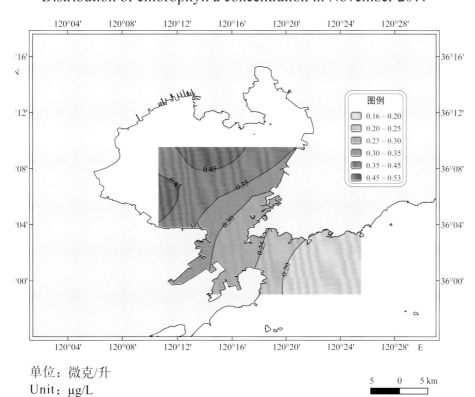

单位：微克/升
Unit：μg/L

2012年2月叶绿素 a 浓度分布
Distribution of chlorophyll a concentration in February 2012

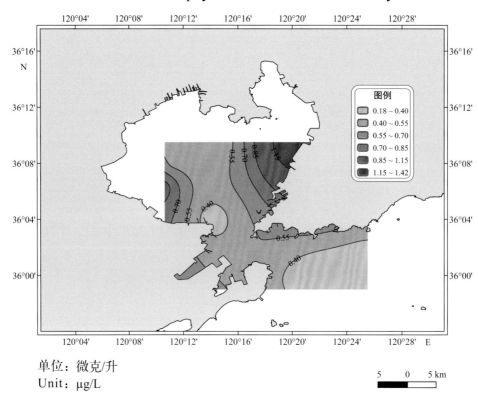

单位：微克/升
Unit：μg/L

2012年5月叶绿素 a 浓度分布
Distribution of chlorophyll a concentration in May 2012

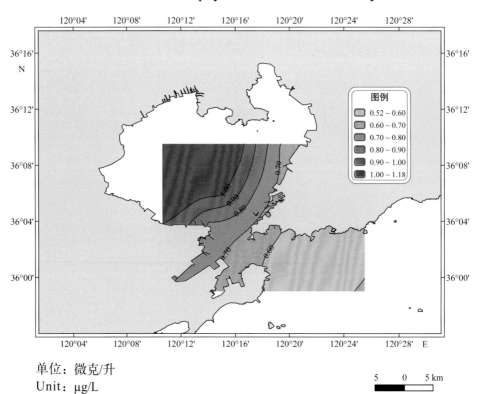

单位：微克/升
Unit：μg/L

2012年8月叶绿素 a 浓度分布
Distribution of chlorophyll a concentration in August 2012

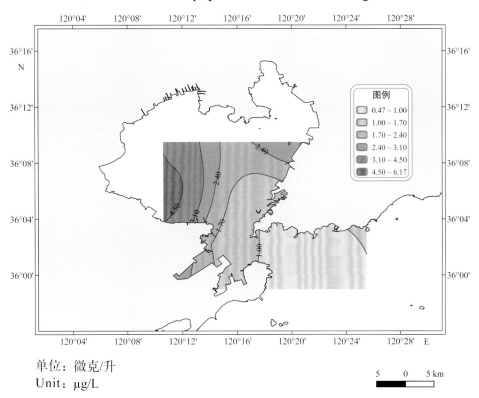

单位：微克/升
Unit：μg/L

2012年11月叶绿素 a 浓度分布
Distribution of chlorophyll a concentration in November 2012

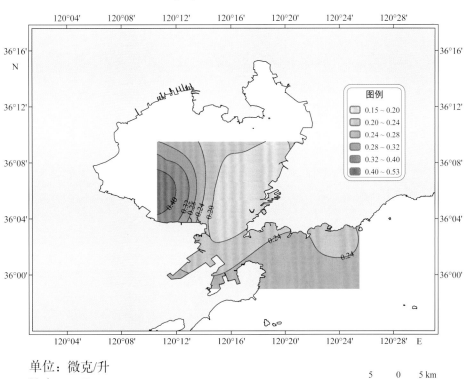

单位：微克/升
Unit：μg/L

2013年2月叶绿素 a 浓度分布
Distribution of chlorophyll a concentration in February 2013

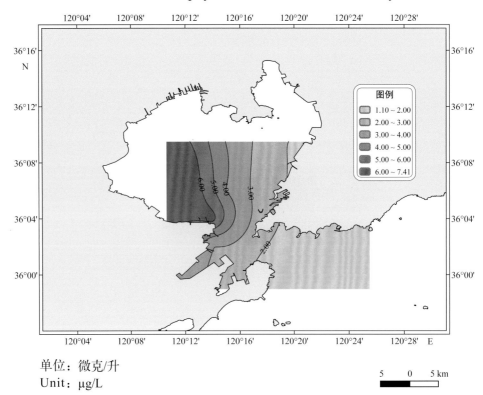

单位：微克/升
Unit：μg/L

2013年5月叶绿素 a 浓度分布
Distribution of chlorophyll a concentration in May 2013

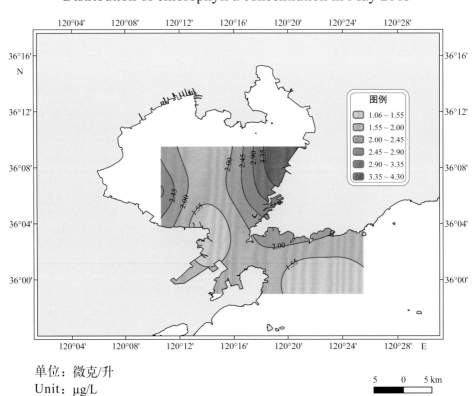

单位：微克/升
Unit：μg/L

2013年8月叶绿素 a 浓度分布
Distribution of chlorophyll a concentration in August 2013

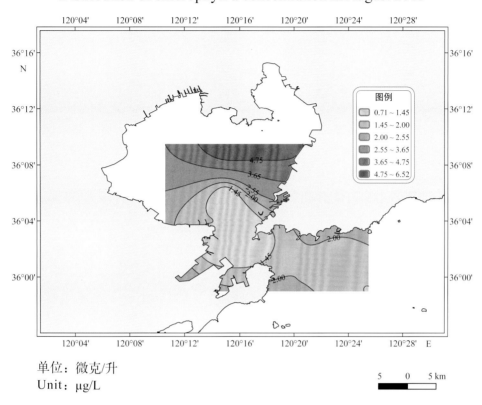

单位：微克/升
Unit：μg/L

2013年11月叶绿素 a 浓度分布
Distribution of chlorophyll a concentration in November 2013

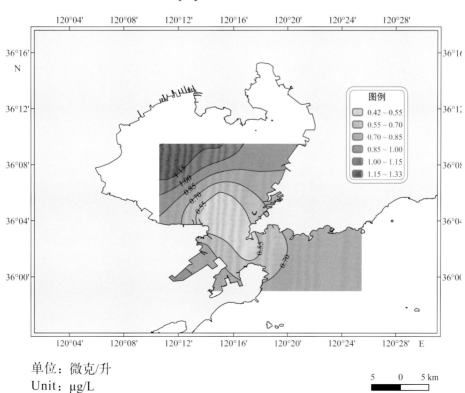

单位：微克/升
Unit：μg/L

2014年2月叶绿素 a 浓度分布
Distribution of chlorophyll a concentration in February 2014

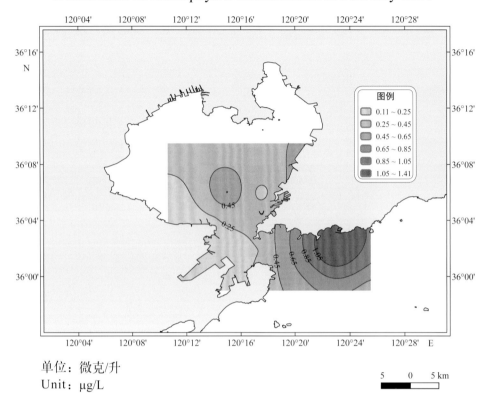

单位：微克/升
Unit：μg/L

2014年5月叶绿素 a 浓度分布
Distribution of chlorophyll a concentration in May 2014

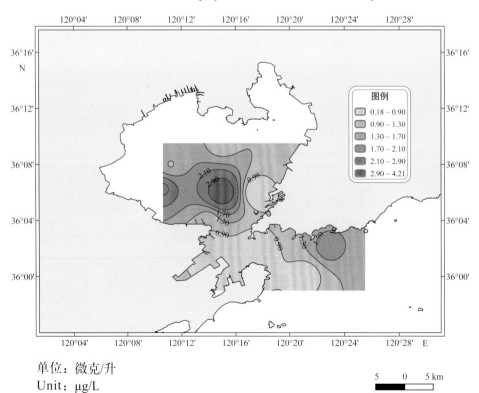

单位：微克/升
Unit：μg/L

2014年8月叶绿素 a 浓度分布
Distribution of chlorophyll a concentration in August 2014

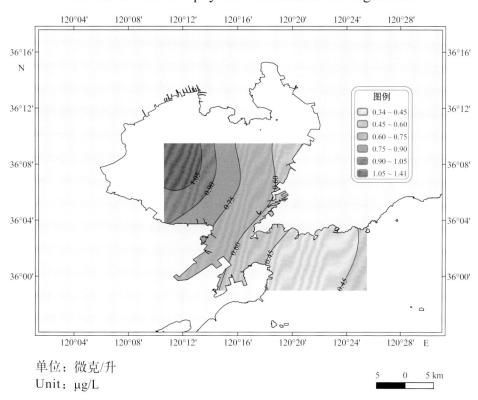

单位：微克/升
Unit：μg/L

2014年11月叶绿素 a 浓度分布
Distribution of chlorophyll a concentration in November 2014

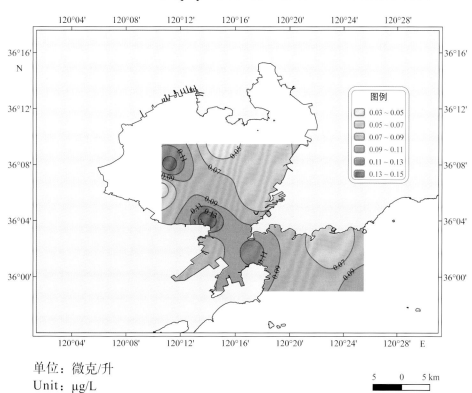

单位：微克/升
Unit：μg/L

2015年2月叶绿素 a 浓度分布
Distribution of chlorophyll a concentration in February 2015

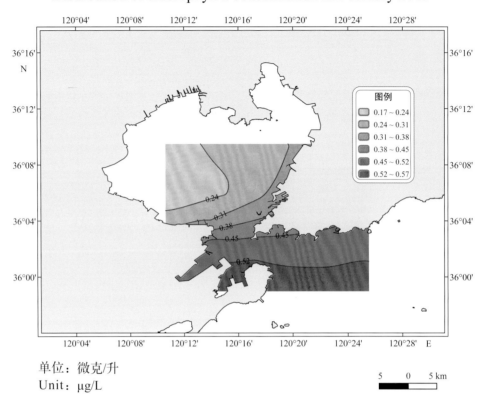

单位：微克/升
Unit：μg/L

2015年5月叶绿素 a 浓度分布
Distribution of chlorophyll a concentration in May 2015

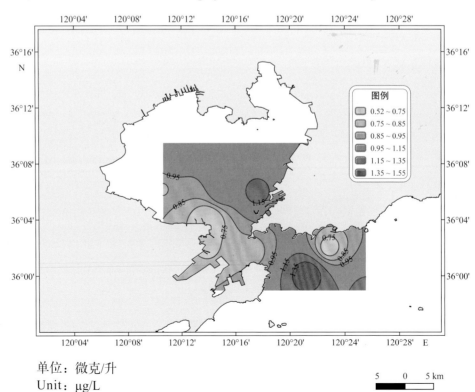

单位：微克/升
Unit：μg/L

2015年8月叶绿素 a 浓度分布
Distribution of chlorophyll a concentration in August 2015

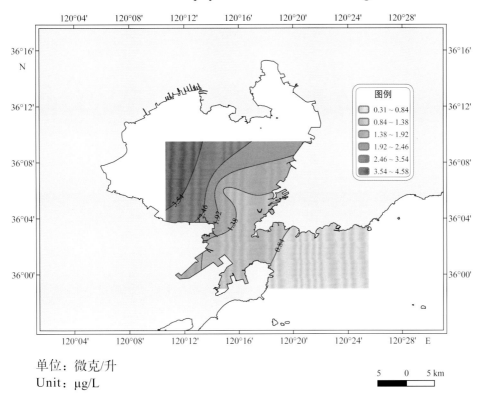

单位：微克/升
Unit：μg/L

2015年11月叶绿素 a 浓度分布
Distribution of chlorophyll a concentration in November 2015

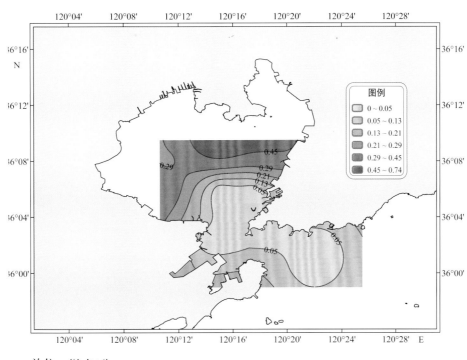

单位：微克/升
Unit：μg/L

2. 浮游植物细胞数量 Cell density of phytoplankton

2010年2月浮游植物细胞数量分布
Distribution of phytoplankton cell density in February 2010

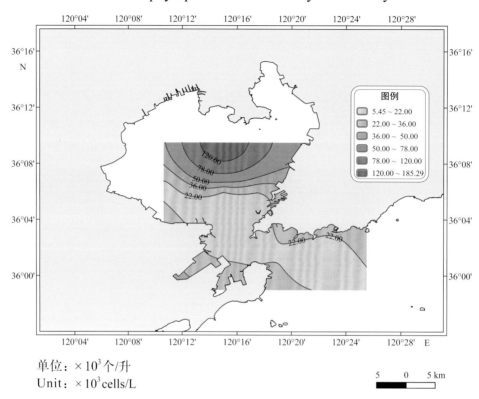

单位：×10³个/升
Unit：×10³cells/L

2010年5月浮游植物细胞数量分布
Distribution of phytoplankton cell density in May 2010

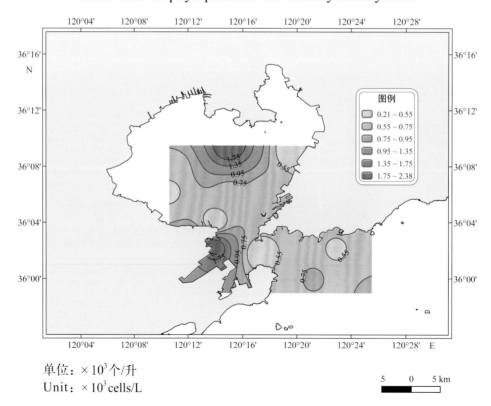

单位：×10³个/升
Unit：×10³cells/L

2010年8月浮游植物细胞数量分布
Distribution of phytoplankton cell density in August 2010

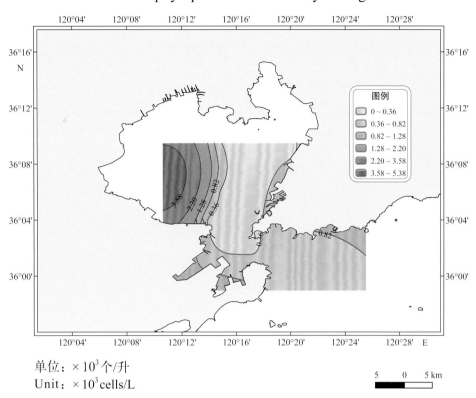

单位：×10³个/升
Unit：×10³cells/L

2010年11月浮游植物细胞数量分布
Distribution of phytoplankton cell density in November 2010

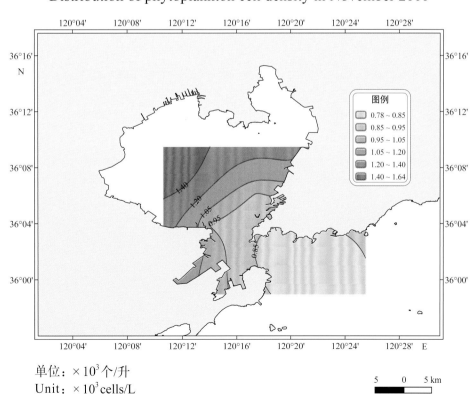

单位：×10³个/升
Unit：×10³cells/L

2011年2月浮游植物细胞数量分布
Distribution of phytoplankton cell density in February 2011

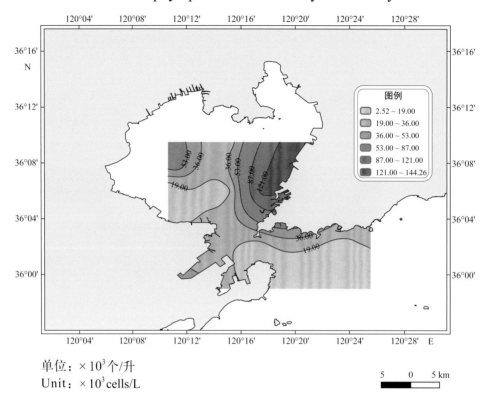

单位：×10³个/升
Unit：×10³cells/L

2011年5月浮游植物细胞数量分布
Distribution of phytoplankton cell density in May 2011

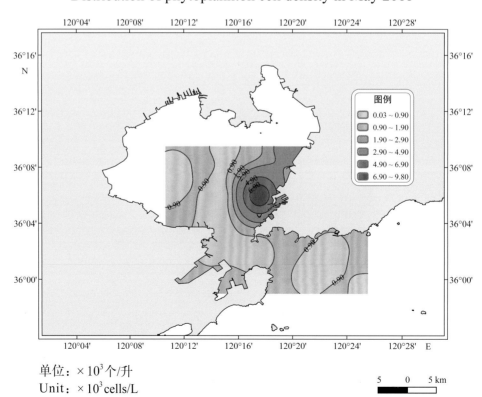

单位：×10³个/升
Unit：×10³cells/L

2011年8月浮游植物细胞数量分布
Distribution of phytoplankton cell density in August 2011

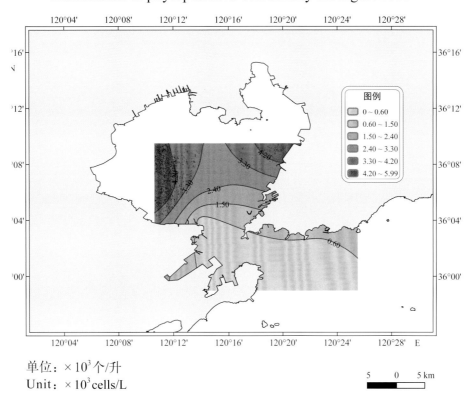

单位：×10³个/升
Unit：×10³cells/L

2011年11月浮游植物细胞数量分布
Distribution of phytoplankton cell density in November 2011

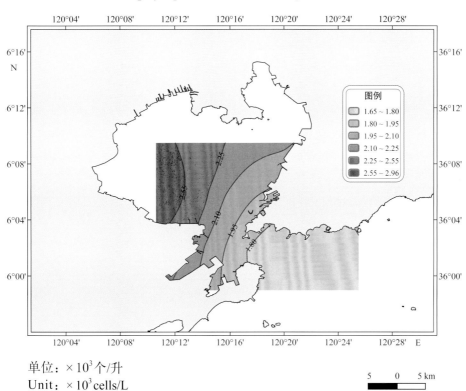

单位：×10³个/升
Unit：×10³cells/L

2012年2月浮游植物细胞数量分布
Distribution of phytoplankton cell density in February 2012

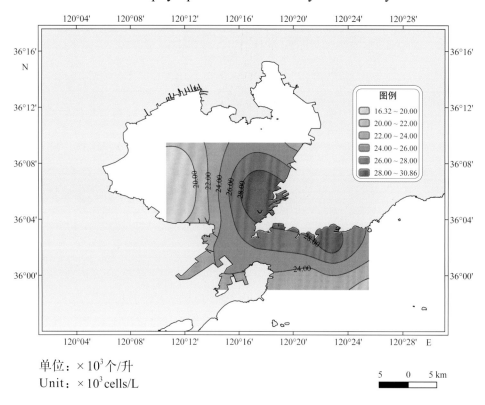

单位：×10³个/升
Unit：×10³ cells/L

2012年5月浮游植物细胞数量分布
Distribution of phytoplankton cell density in May 2012

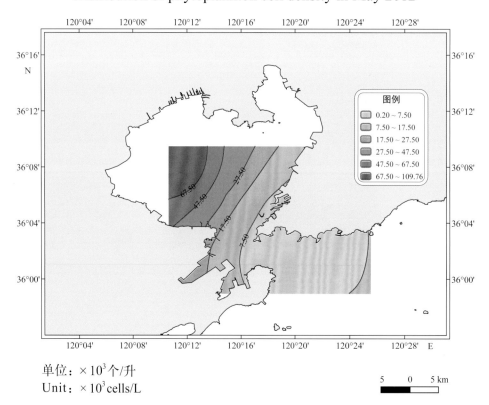

单位：×10³个/升
Unit：×10³ cells/L

2012年8月浮游植物细胞数量分布
Distribution of phytoplankton cell density in August 2012

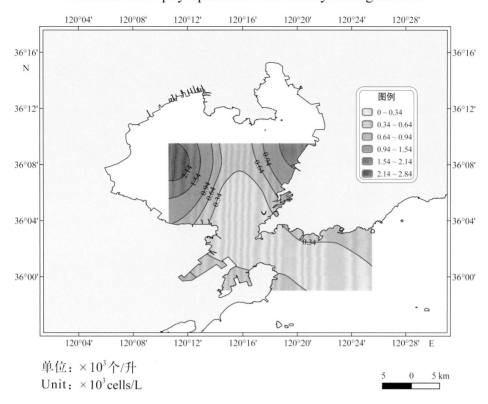

单位：×10³个/升
Unit：×10³cells/L

2012年11月浮游植物细胞数量分布
Distribution of phytoplankton cell density in November 2012

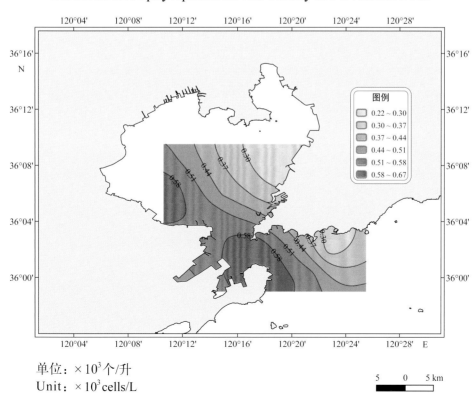

单位：×10³个/升
Unit：×10³cells/L

2013年2月浮游植物细胞数量分布
Distribution of phytoplankton cell density in February 2013

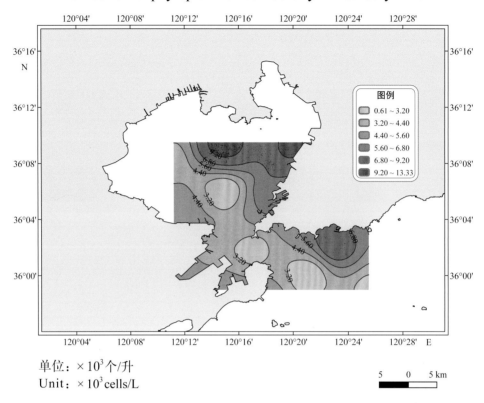

单位：×10³个/升
Unit：×10³ cells/L

2013年5月浮游植物细胞数量分布
Distribution of phytoplankton cell density in May 2013

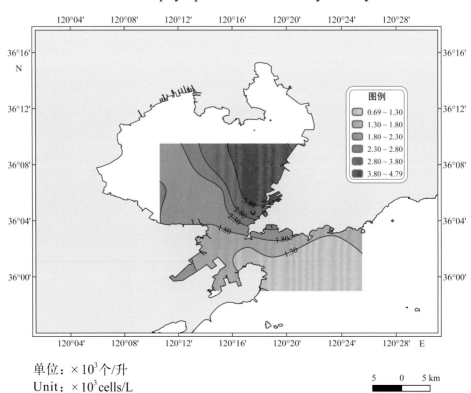

单位：×10³个/升
Unit：×10³ cells/L

2013年8月浮游植物细胞数量分布
Distribution of phytoplankton cell density in August 2013

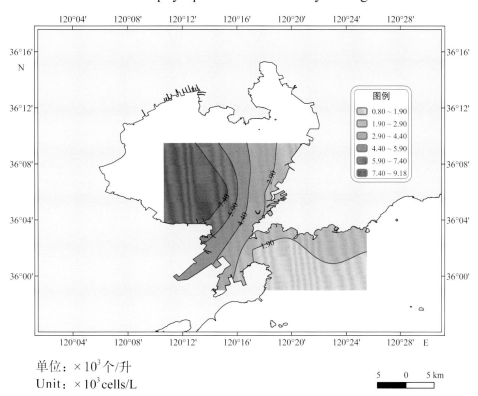

单位：×10³个/升
Unit：×10³ cells/L

2013年11月浮游植物细胞数量分布
Distribution of phytoplankton cell density in November 2013

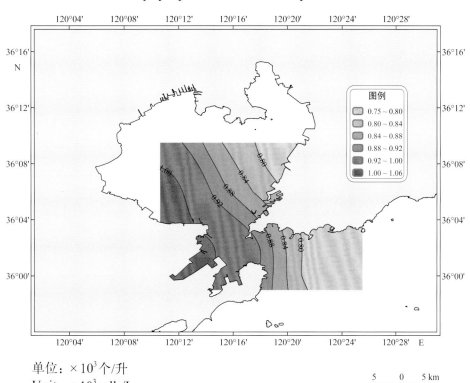

单位：×10³个/升
Unit：×10³ cells/L

2014年2月浮游植物细胞数量分布
Distribution of phytoplankton cell density in February 2014

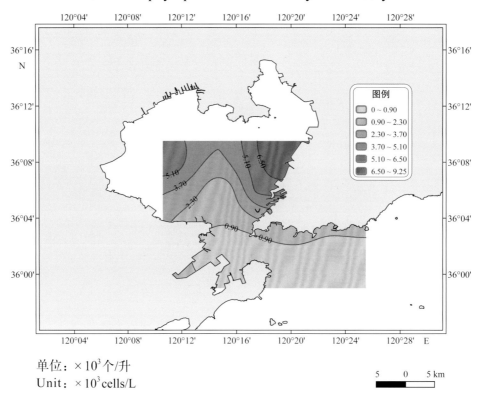

单位：×10³个/升
Unit：×10³ cells/L

2014年5月浮游植物细胞数量分布
Distribution of phytoplankton cell density in May 2014

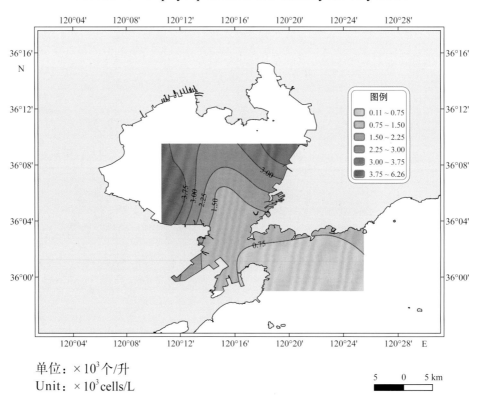

单位：×10³个/升
Unit：×10³ cells/L

2014年8月浮游植物细胞数量分布
Distribution of phytoplankton cell density in August 2014

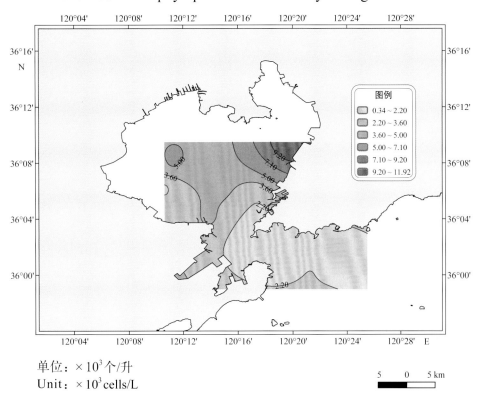

单位：×10³个/升
Unit：×10³cells/L

2014年11月浮游植物细胞数量分布
Distribution of phytoplankton cell density in November 2014

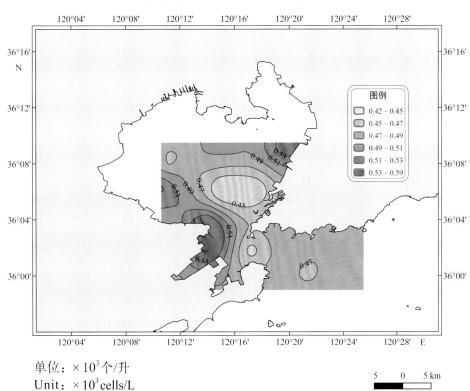

单位：×10³个/升
Unit：×10³cells/L

2015年2月浮游植物细胞数量分布
Distribution of phytoplankton cell density in February 2015

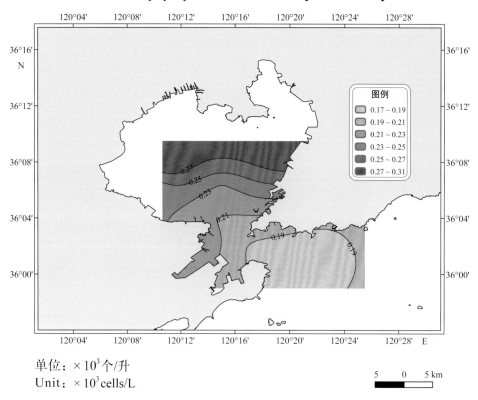

单位：×10³个/升
Unit：×10³ cells/L

2015年5月浮游植物细胞数量分布
Distribution of phytoplankton cell density in May 2015

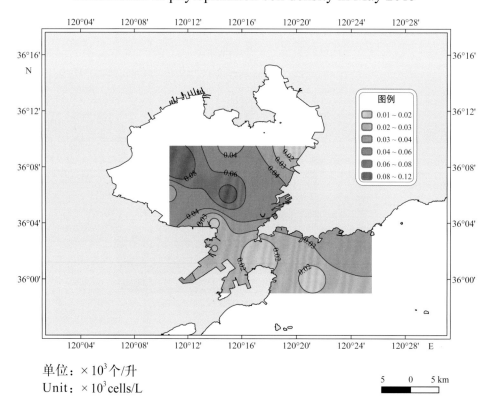

单位：×10³个/升
Unit：×10³ cells/L

2015年8月浮游植物细胞数量分布
Distribution of phytoplankton cell density in August 2015

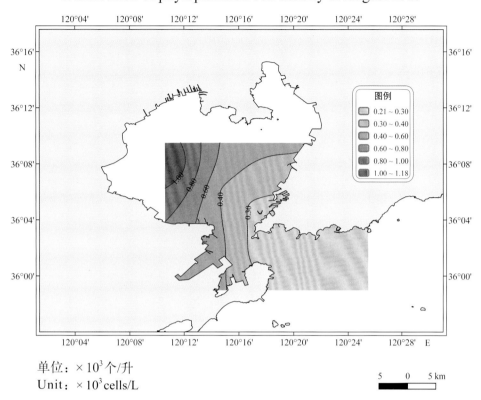

单位：×10³个/升
Unit：×10³cells/L

2015年11月浮游植物细胞数量分布
Distribution of phytoplankton cell density in November 2015

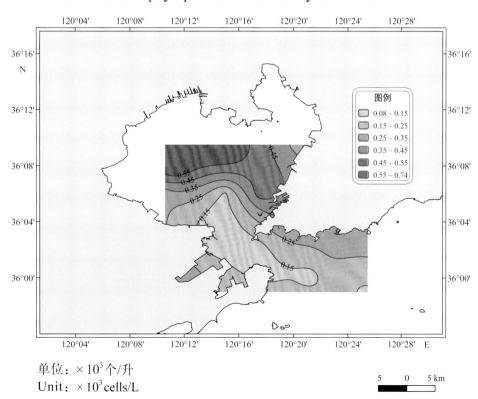

单位：×10³个/升
Unit：×10³cells/L

3. 浮游动物丰度 Abundance of zooplankton

2010年2月浮游动物丰度分布
Distribution of zooplankton abundance in February 2010

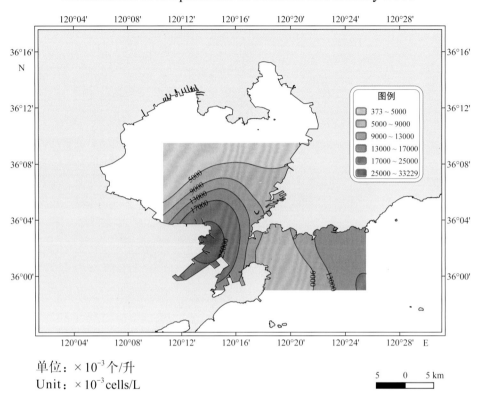

单位：$\times 10^{-3}$ 个/升
Unit：$\times 10^{-3}$ cells/L

2010年5月浮游动物丰度分布
Distribution of zooplankton abundance in May 2010

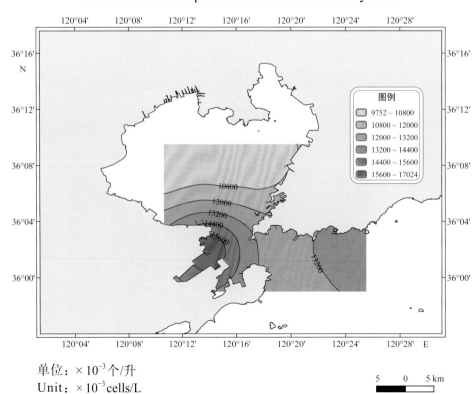

单位：$\times 10^{-3}$ 个/升
Unit：$\times 10^{-3}$ cells/L

2010年8月浮游动物丰度分布
Distribution of zooplankton abundance in August 2010

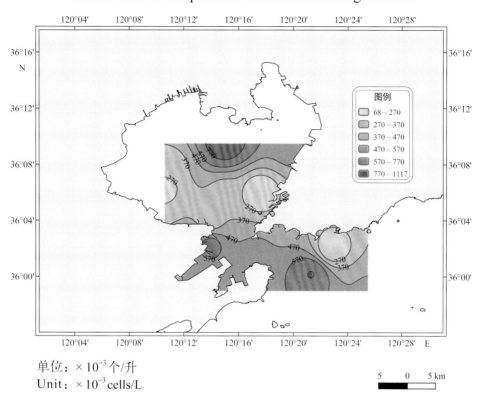

单位：×10⁻³ 个/升
Unit：$\times 10^{-3}$ cells/L

2010年11月浮游动物丰度分布
Distribution of zooplankton abundance in November 2010

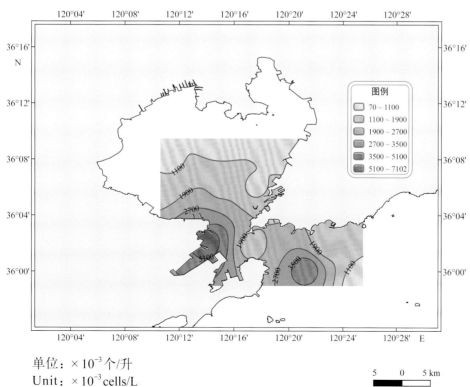

单位：×10⁻³ 个/升
Unit：$\times 10^{-3}$ cells/L

2011年2月浮游动物丰度分布
Distribution of zooplankton abundance in February 2011

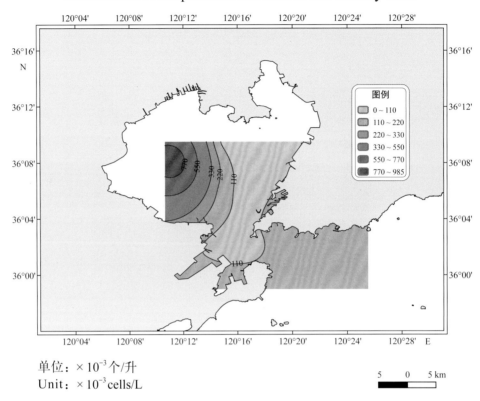

单位：$\times 10^{-3}$ 个/升
Unit：$\times 10^{-3}$ cells/L

2011年5月浮游动物丰度分布
Distribution of zooplankton abundance in May 2011

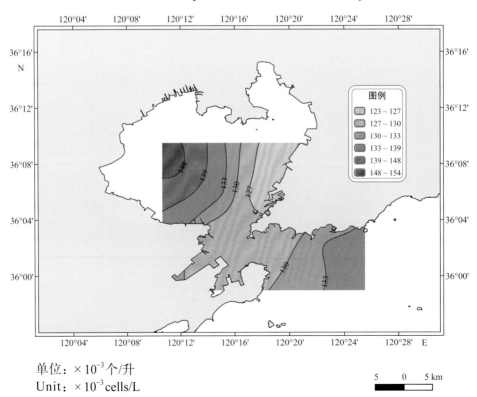

单位：$\times 10^{-3}$ 个/升
Unit：$\times 10^{-3}$ cells/L

2011年8月浮游动物丰度分布
Distribution of zooplankton abundance in August 2011

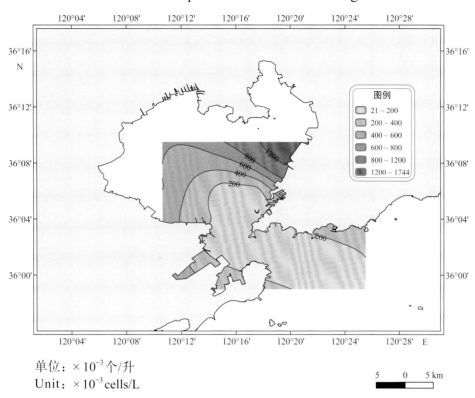

单位：×10^{-3}个/升
Unit：×10^{-3}cells/L

2011年11月浮游动物丰度分布
Distribution of zooplankton abundance in November 2011

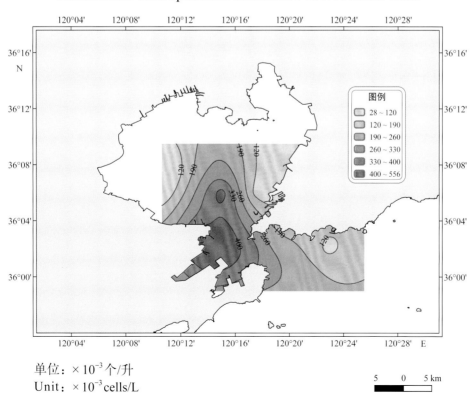

单位：×10^{-3}个/升
Unit：×10^{-3}cells/L

2012年2月浮游动物丰度分布
Distribution of zooplankton abundance in February 2012

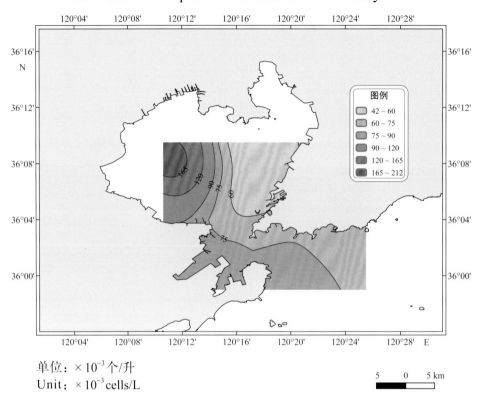

单位：×10^{-3} 个/升
Unit：×10^{-3} cells/L

2012年5月浮游动物丰度分布
Distribution of zooplankton abundance in May 2012

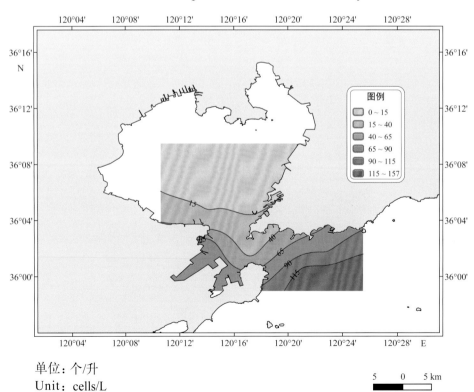

单位：个/升
Unit：cells/L

2012年8月浮游动物丰度分布
Distribution of zooplankton abundance in August 2012

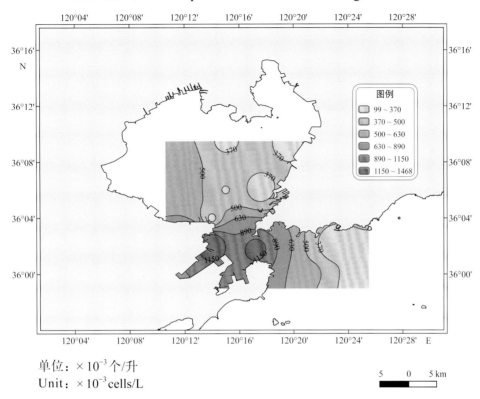

单位：×10^{-3} 个/升
Unit：×10^{-3} cells/L

2012年11月浮游动物丰度分布
Distribution of zooplankton abundance in November 2012

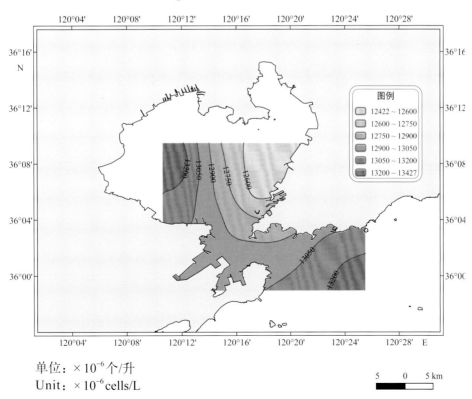

单位：×10^{-6} 个/升
Unit：×10^{-6} cells/L

2013年2月浮游动物丰度分布
Distribution of zooplankton abundance in February 2013

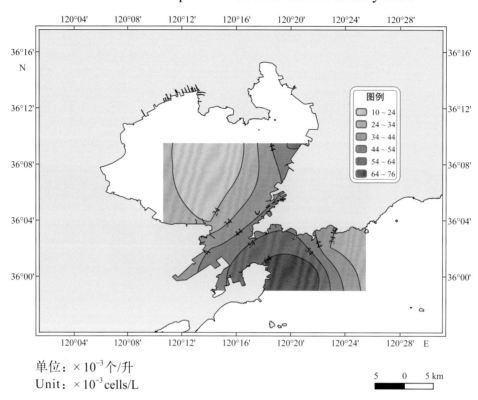

单位：×10⁻³ 个/升
Unit: $\times 10^{-3}$ cells/L

2013年5月浮游动物丰度分布
Distribution of zooplankton abundance in May 2013

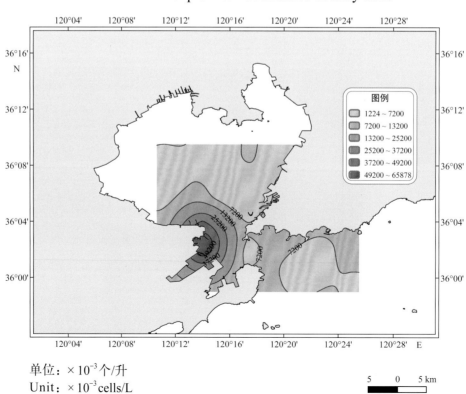

单位：×10⁻³ 个/升
Unit: $\times 10^{-3}$ cells/L

2013年8月浮游动物丰度分布
Distribution of zooplankton abundance in August 2013

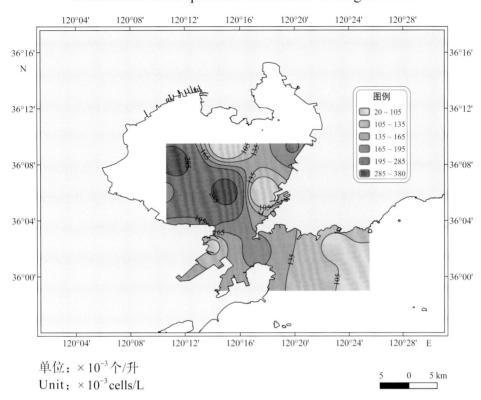

单位：$\times 10^{-3}$ 个/升
Unit：$\times 10^{-3}$ cells/L

2013年11月浮游动物丰度分布
Distribution of zooplankton abundance in November 2013

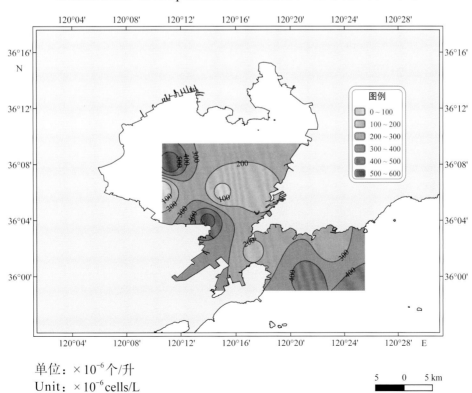

单位：$\times 10^{-6}$ 个/升
Unit：$\times 10^{-6}$ cells/L

2014年2月浮游动物丰度分布
Distribution of zooplankton abundance in February 2014

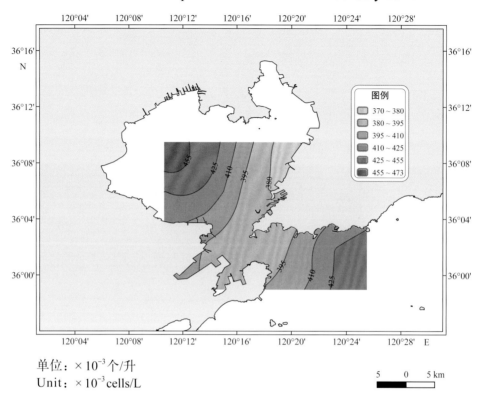

单位：×10^{-3}个/升
Unit：×10^{-3}cells/L

2014年5月浮游动物丰度分布
Distribution of zooplankton abundance in May 2014

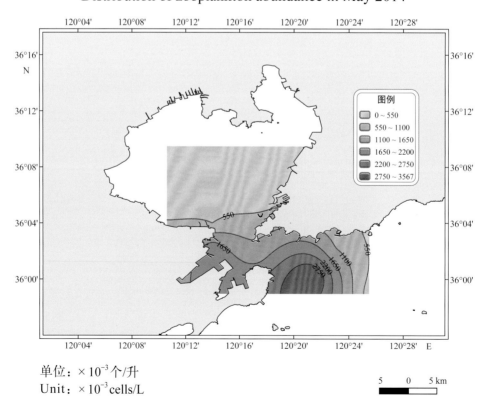

单位：×10^{-3}个/升
Unit：×10^{-3}cells/L

2014年8月浮游动物丰度分布
Distribution of zooplankton abundance in August 2014

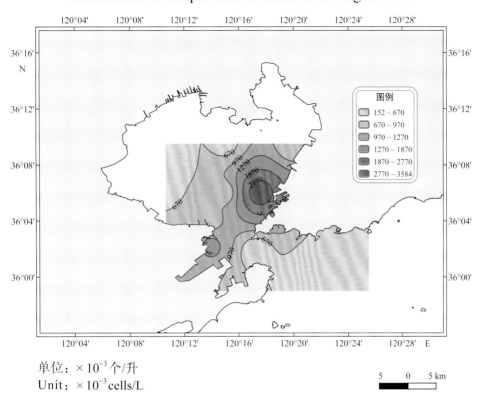

单位：$\times 10^{-3}$ 个/升
Unit：$\times 10^{-3}$ cells/L

2014年11月浮游动物丰度分布
Distribution of zooplankton abundance in November 2014

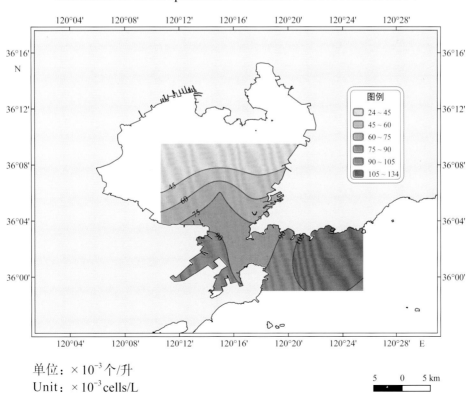

单位：$\times 10^{-3}$ 个/升
Unit：$\times 10^{-3}$ cells/L

2015年2月浮游动物丰度分布
Distribution of zooplankton abundance in February 2015

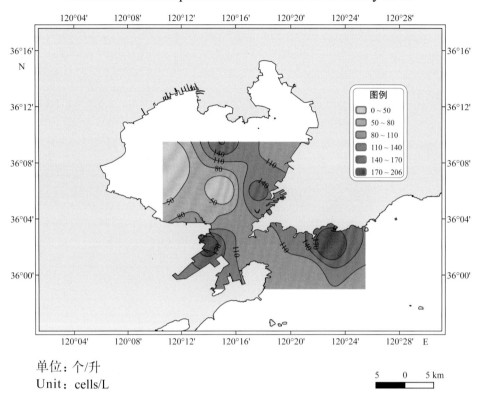

单位：个/升
Unit: cells/L

2015年5月浮游动物丰度分布
Distribution of zooplankton abundance in May 2015

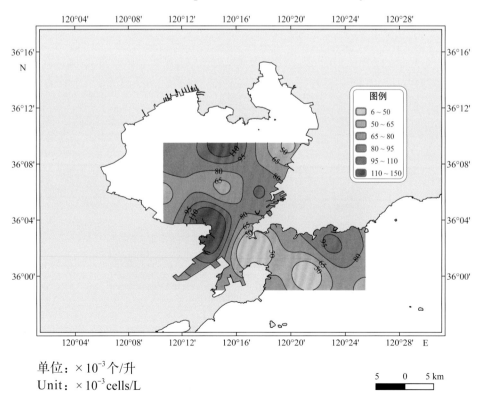

单位：$\times 10^{-3}$ 个/升
Unit: $\times 10^{-3}$ cells/L

2015年8月浮游动物丰度分布
Distribution of zooplankton abundance in August 2015

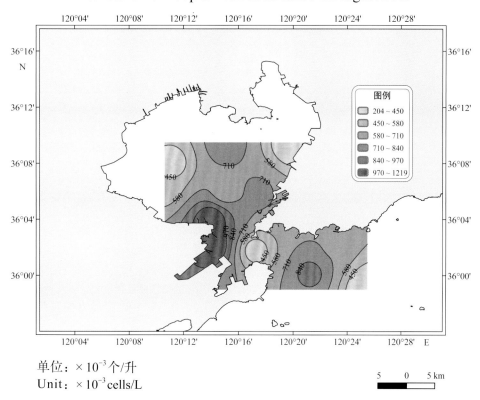

单位：×10⁻³个/升
Unit：$\times 10^{-3}$ cells/L

2015年11月浮游动物丰度分布
Distribution of zooplankton abundance in November 2015

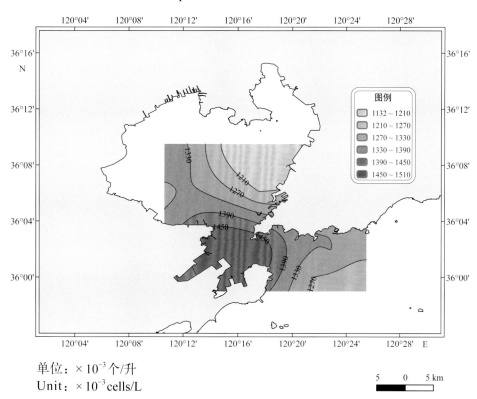

单位：×10⁻³个/升
Unit：$\times 10^{-3}$ cells/L

4. 浮游动物生物量 Biomass of zooplankton

2010年2月浮游动物生物量分布
Distribution of zooplankton biomass in February 2010

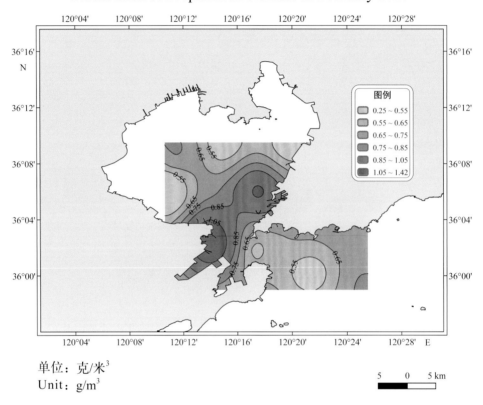

单位：克/米³
Unit：g/m³

2010年5月浮游动物生物量分布
Distribution of zooplankton biomass in May 2010

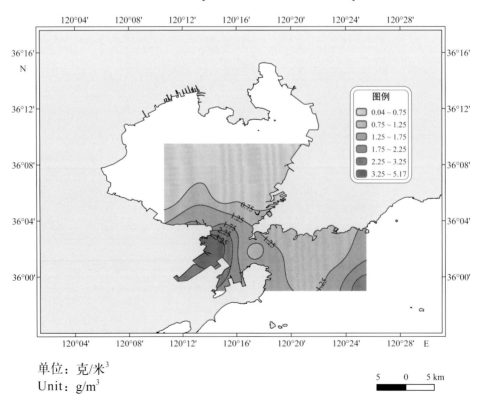

单位：克/米³
Unit：g/m³

2010年8月浮游动物生物量分布
Distribution of zooplankton biomass in August 2010

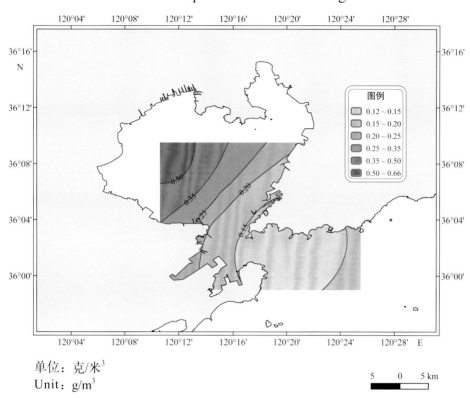

单位：克/米³
Unit：g/m³

2010年11月浮游动物生物量分布
Distribution of zooplankton biomass in November 2010

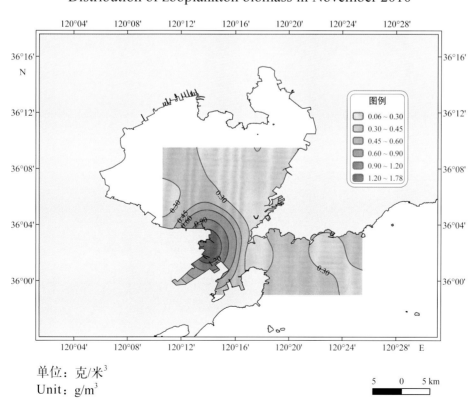

单位：克/米³
Unit：g/m³

2011年2月浮游动物生物量分布
Distribution of zooplankton biomass in February 2011

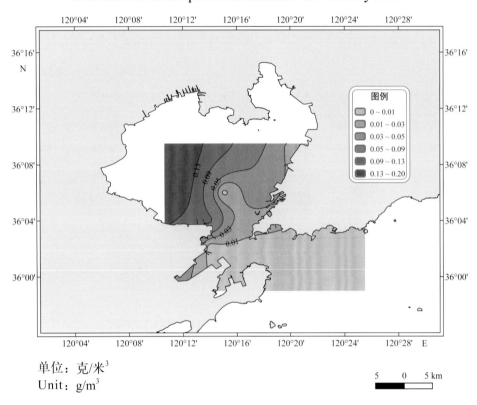

单位：克/米³
Unit：g/m³

2011年5月浮游动物生物量分布
Distribution of zooplankton biomass in May 2011

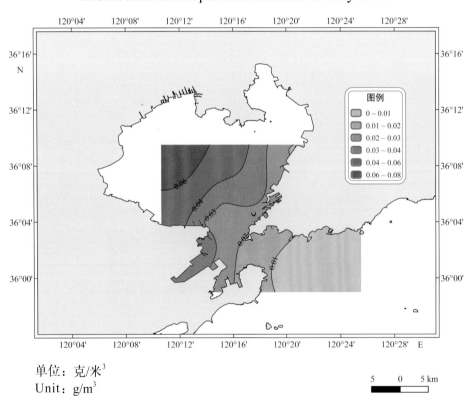

单位：克/米³
Unit：g/m³

2011年8月浮游动物生物量分布
Distribution of zooplankton biomass in August 2011

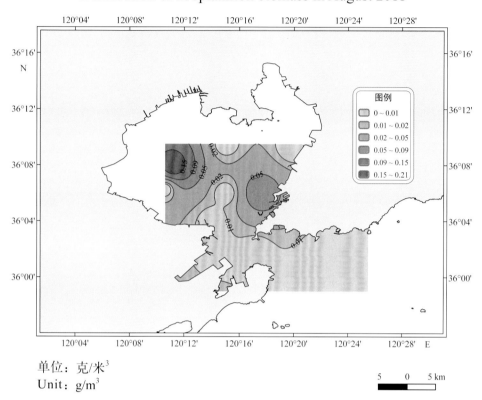

单位：克/米³
Unit：g/m³

2011年11月浮游动物生物量分布
Distribution of zooplankton biomass in November 2011

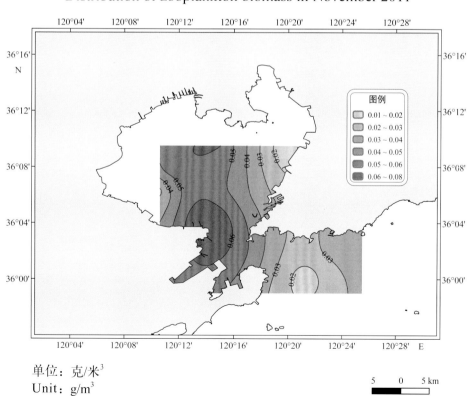

单位：克/米³
Unit：g/m³

2012年2月浮游动物生物量分布
Distribution of zooplankton biomass in February 2012

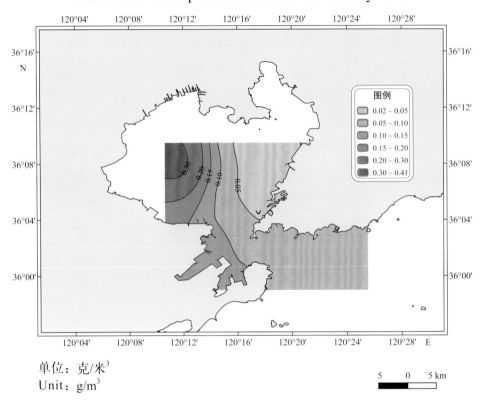

单位：克/米³
Unit：g/m³

2012年5月浮游动物生物量分布
Distribution of zooplankton biomass in May 2012

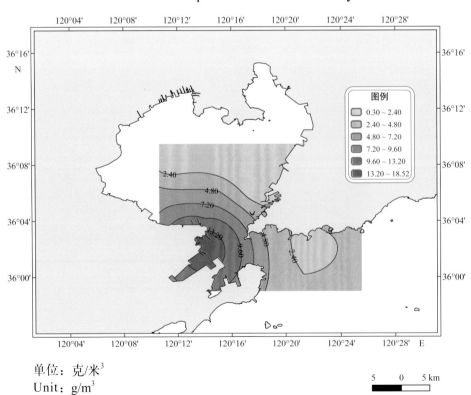

单位：克/米³
Unit：g/m³

2012年8月浮游动物生物量分布
Distribution of zooplankton biomass in August 2012

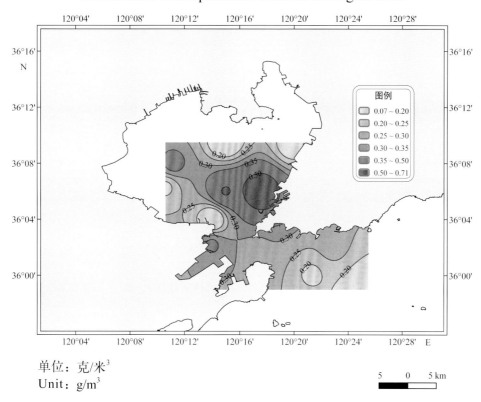

单位：克/米^3
Unit：g/m^3

2012年11月浮游动物生物量分布
Distribution of zooplankton biomass in November 2012

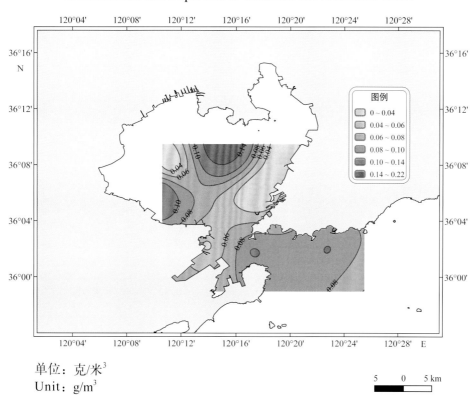

单位：克/米^3
Unit：g/m^3

2013年2月浮游动物生物量分布
Distribution of zooplankton biomass in February 2013

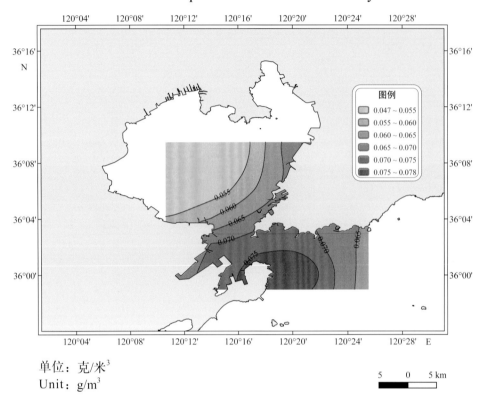

单位：克/米³
Unit：g/m³

2013年5月浮游动物生物量分布
Distribution of zooplankton biomass in May 2013

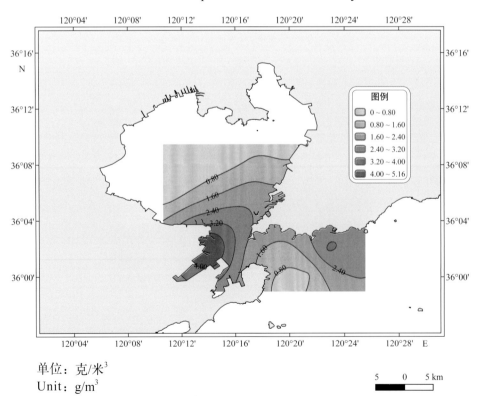

单位：克/米³
Unit：g/m³

2013年8月浮游动物生物量分布
Distribution of zooplankton biomass in August 2013

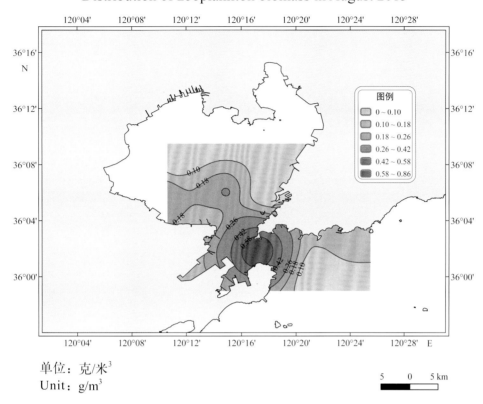

单位：克/米^3
Unit: g/m^3

2013年11月浮游动物生物量分布
Distribution of zooplankton biomass in November 2013

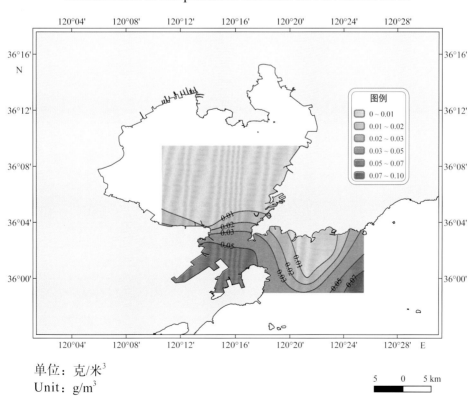

单位：克/米^3
Unit: g/m^3

2014年2月浮游动物生物量分布
Distribution of zooplankton biomass in February 2014

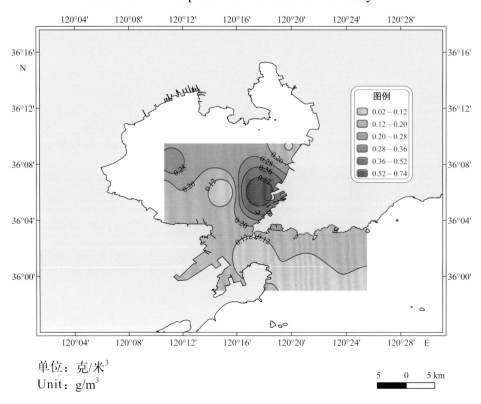

单位：克/米³
Unit：g/m³

2014年5月浮游动物生物量分布
Distribution of zooplankton biomass in May 2014

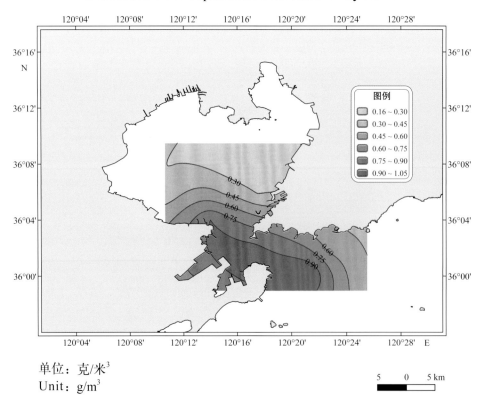

单位：克/米³
Unit：g/m³

2014年8月浮游动物生物量分布
Distribution of zooplankton biomass in August 2014

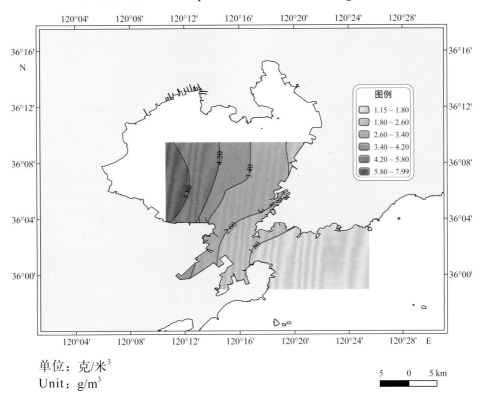

单位：克/米³
Unit：g/m³

2014年11月浮游动物生物量分布
Distribution of zooplankton biomass in November 2014

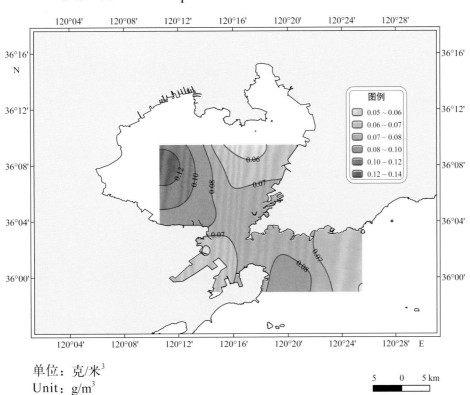

单位：克/米³
Unit：g/m³

2015年2月浮游动物生物量分布
Distribution of zooplankton biomass in February 2015

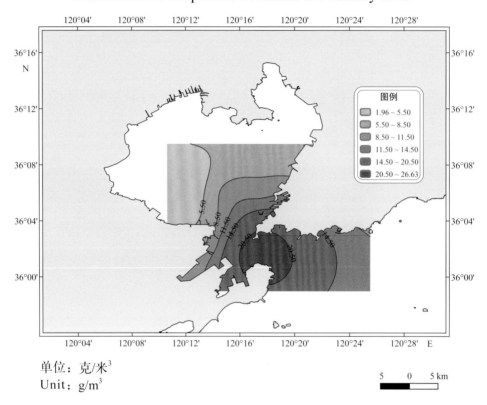

单位:克/米^3
Unit:g/m^3

2015年5月浮游动物生物量分布
Distribution of zooplankton biomass in May 2015

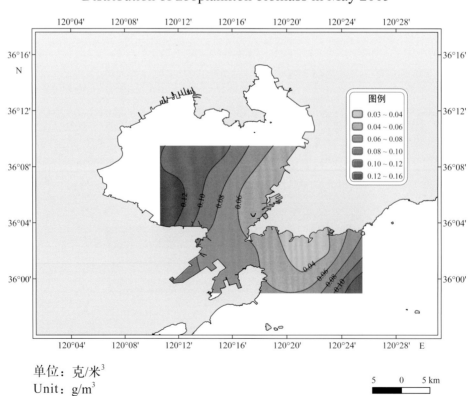

单位:克/米^3
Unit:g/m^3

2015年8月浮游动物生物量分布
Distribution of zooplankton biomass in August 2015

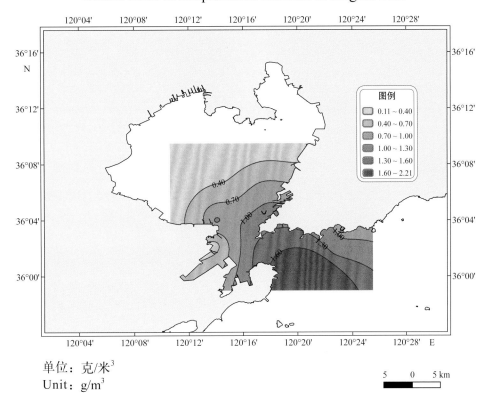

单位：克/米³
Unit：g/m³

2015年11月浮游动物生物量分布
Distribution of zooplankton biomass in November 2015

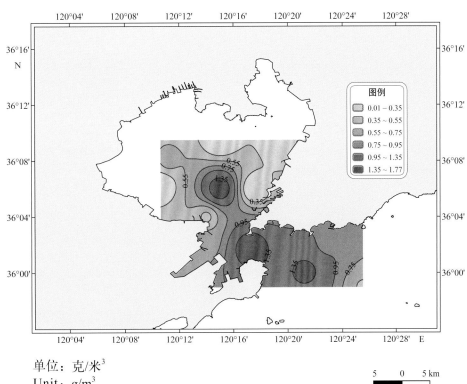

单位：克/米³
Unit：g/m³

5. 浮游动物优势种组成 Dominant species composition of zooplankton

2010年2月浮游动物优势种组成分布
Composition distribution of dominant zooplankton species in February 2010

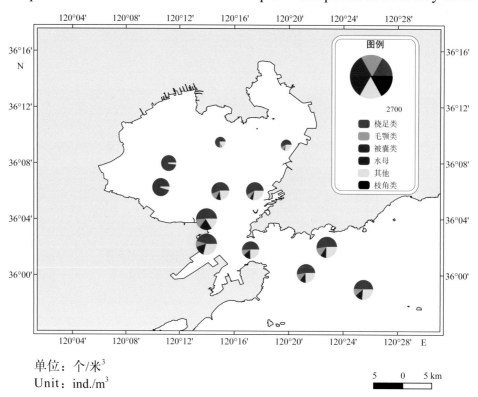

单位：个/米³
Unit：ind./m³

2010年5月浮游动物优势种组成分布
Composition distribution of dominant zooplankton species in May 2010

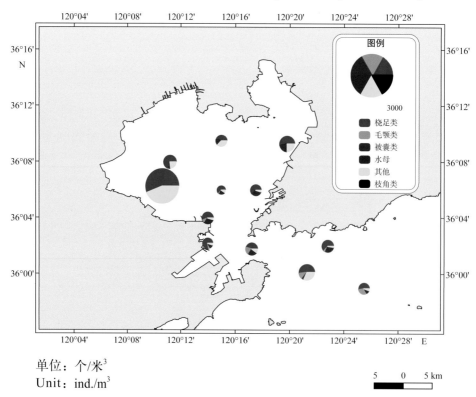

单位：个/米³
Unit：ind./m³

2010年8月浮游动物优势种组成分布
Composition distribution of dominant zooplankton species in August 2010

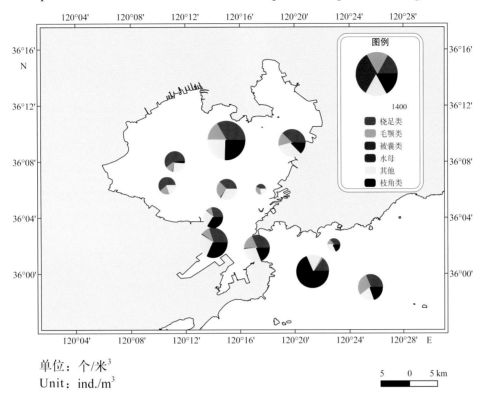

2010年11月浮游动物优势种组成分布
Composition distribution of dominant zooplankton species in November 2010

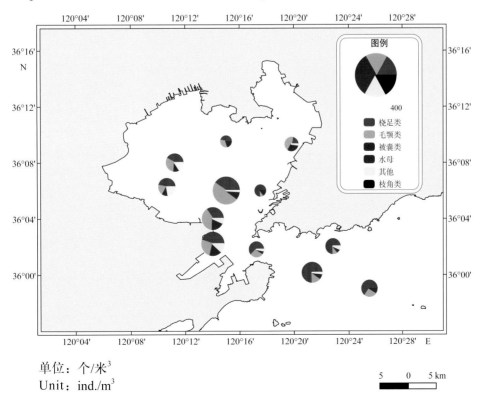

2011年2月浮游动物优势种组成分布
Composition distribution of dominant zooplankton species in February 2011

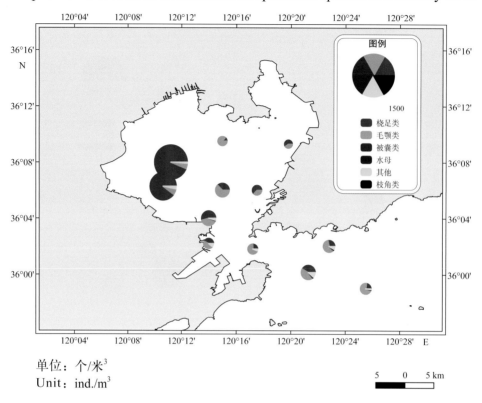

2011年5月浮游动物优势种组成分布
Composition distribution of dominant zooplankton species in May 2011

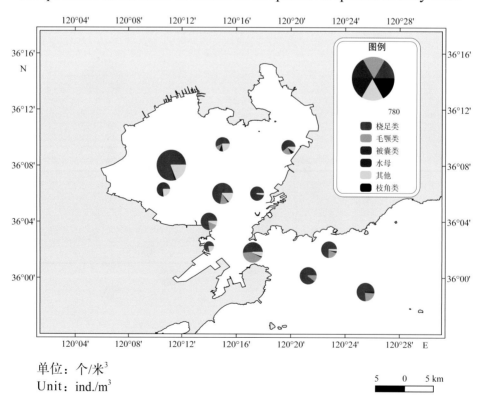

2011年8月浮游动物优势种组成分布
Composition distribution of dominant zooplankton species in August 2011

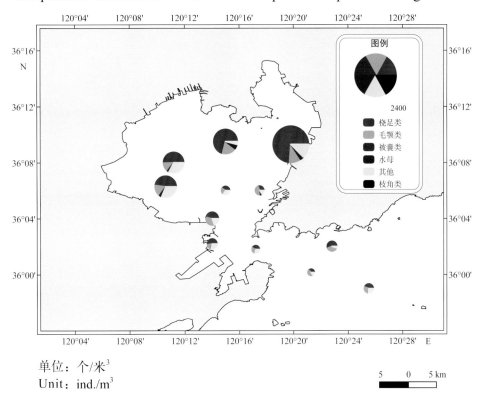

单位：个/米³
Unit: ind./m³

2011年11月浮游动物优势种组成分布
Composition distribution of dominant zooplankton species in November 2011

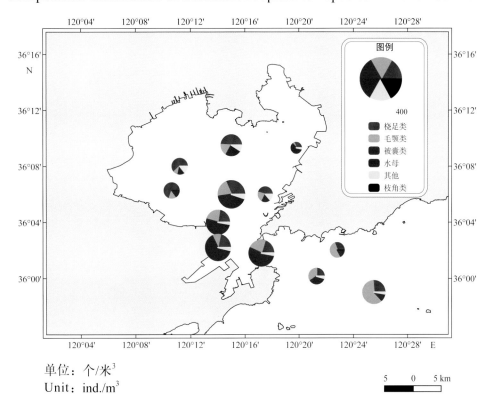

单位：个/米³
Unit: ind./m³

2012年2月浮游动物优势种组成分布
Composition distribution of dominant zooplankton species in February 2012

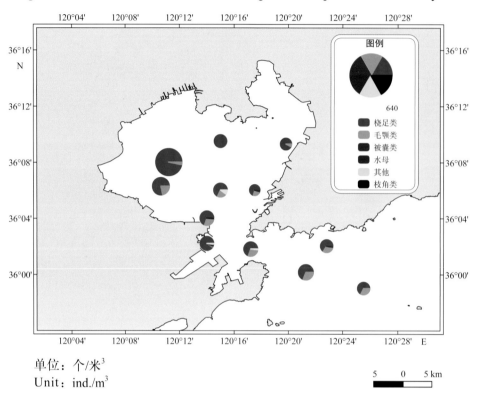

单位：个/米³
Unit：ind./m³

2012年5月浮游动物优势种组成分布
Composition distribution of dominant zooplankton species in May 2012

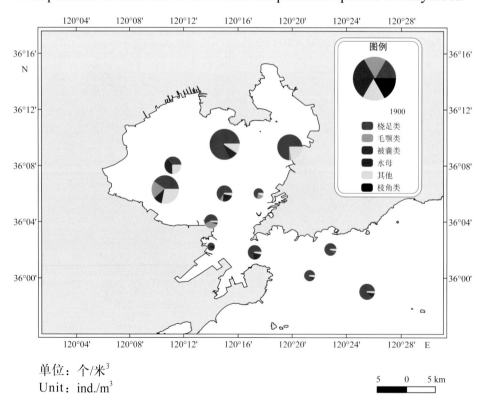

单位：个/米³
Unit：ind./m³

2012年8月浮游动物优势种组成分布
Composition distribution of dominant zooplankton species in August 2012

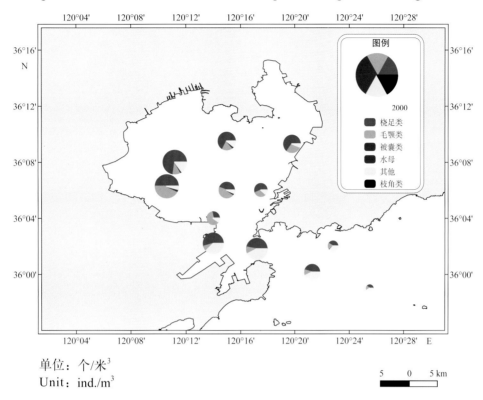

单位：个/米³
Unit：ind./m³

2012年11月浮游动物优势种组成分布
Composition distribution of dominant zooplankton species in November 2012

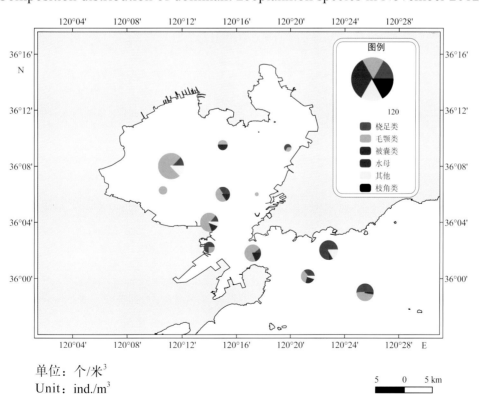

单位：个/米³
Unit：ind./m³

2013年2月浮游动物优势种组成分布
Composition distribution of dominant zooplankton species in February 2013

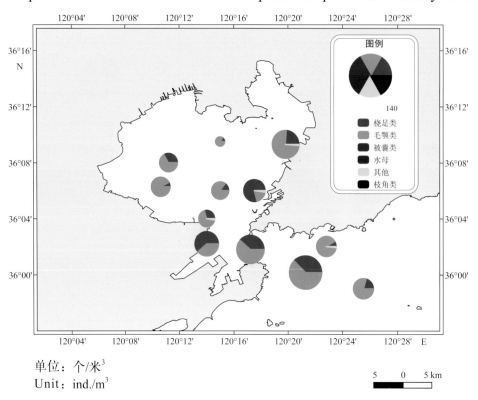

单位：个/米³
Unit: ind./m³

2013年5月浮游动物优势种组成分布
Composition distribution of dominant zooplankton species in May 2013

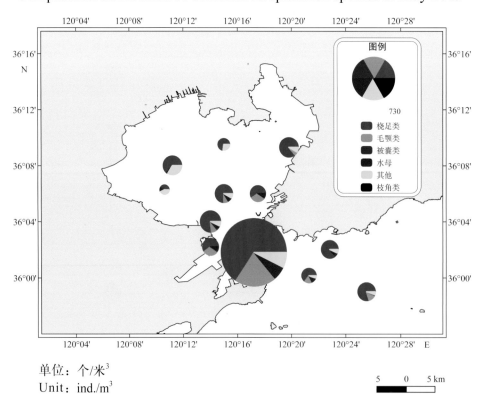

单位：个/米³
Unit: ind./m³

2013年8月浮游动物优势种组成分布
Composition distribution of dominant zooplankton species in August 2013

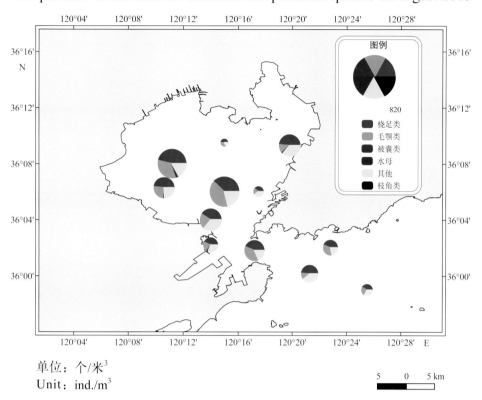

单位：个/米³
Unit: ind./m³

2013年11月浮游动物优势种组成分布
Composition distribution of dominant zooplankton species in November 2013

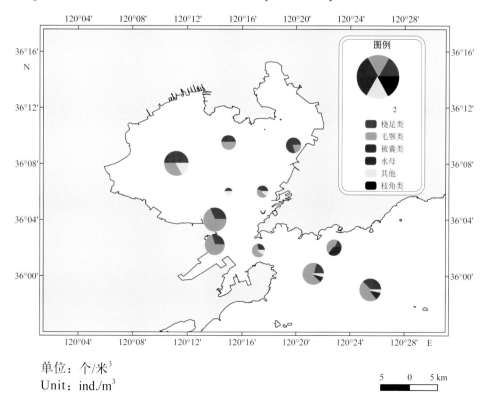

单位：个/米³
Unit: ind./m³

2014年2月浮游动物优势种组成分布
Composition distribution of dominant zooplankton species in February 2014

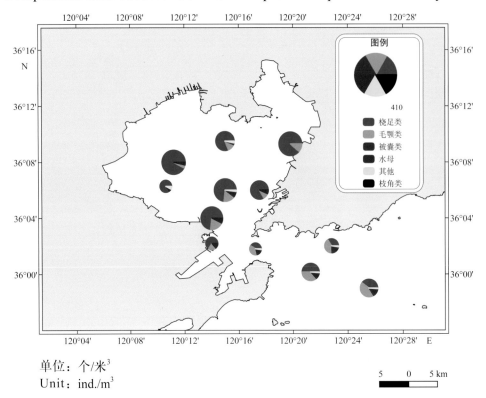

单位：个/米³
Unit：ind./m³

2014年5月浮游动物优势种组成分布
Composition distribution of dominant zooplankton species in May 2014

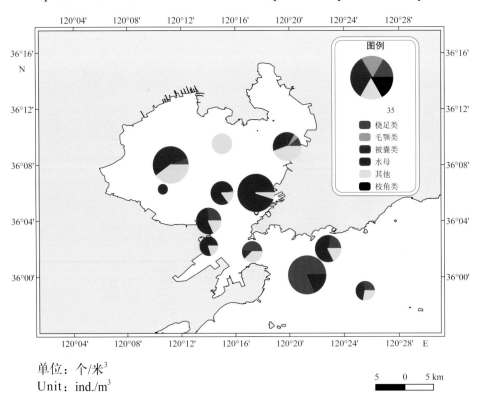

单位：个/米³
Unit：ind./m³

2014年8月浮游动物优势种组成分布
Composition distribution of dominant zooplankton species in August 2014

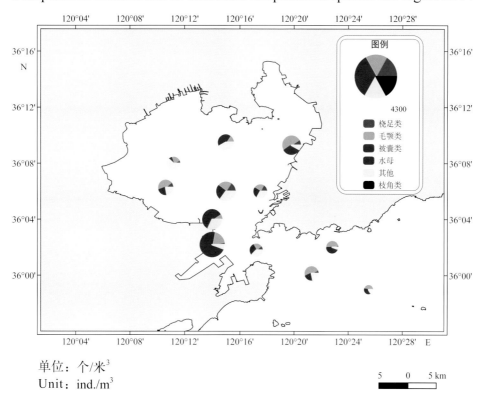

单位：个/米³
Unit：ind./m³

2014年11月浮游动物优势种组成分布
Composition distribution of dominant zooplankton species in November 2014

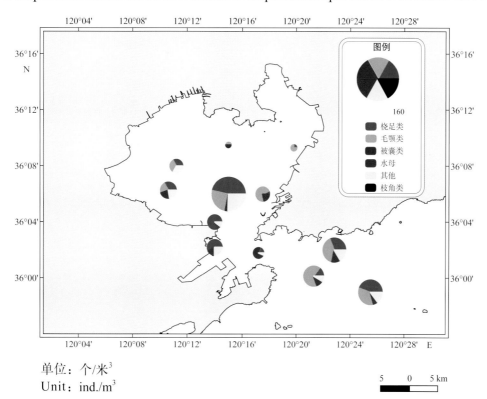

单位：个/米³
Unit：ind./m³

2015年2月浮游动物优势种组成分布
Composition distribution of dominant zooplankton species in February 2015

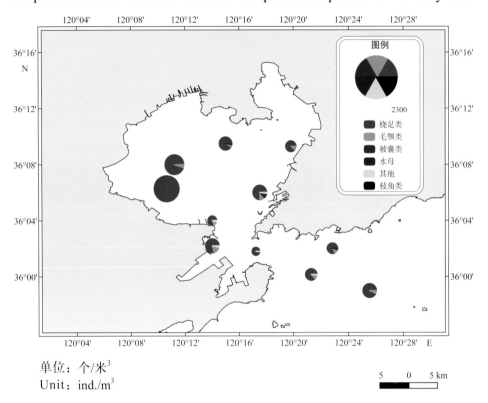

单位：个/米³
Unit：ind./m³

2015年5月浮游动物优势种组成分布
Composition distribution of dominant zooplankton species in May 2015

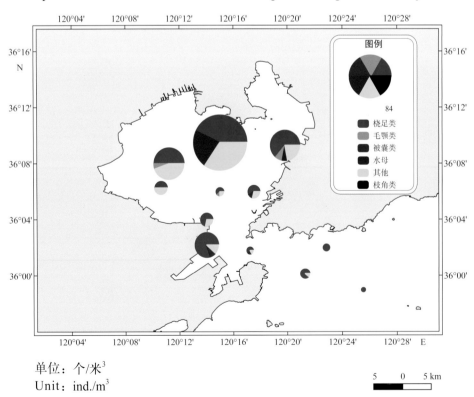

单位：个/米³
Unit：ind./m³

2015年8月浮游动物优势种组成分布
Composition distribution of dominant zooplankton species in August 2015

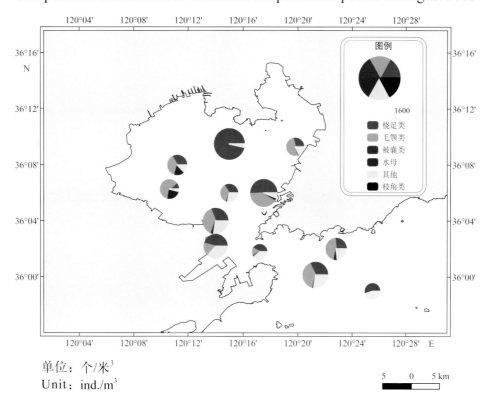

单位：个/米³
Unit: ind./m³

2015年11月浮游动物优势种组成分布
Composition distribution of dominant zooplankton species in November 2015

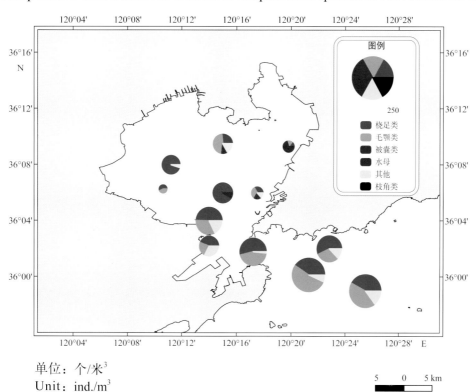

单位：个/米³
Unit: ind./m³

6. 水样含菌数 Bacterial abundance

2010年2月水样含菌数分布
Distribution of bacterial abundance in February 2010

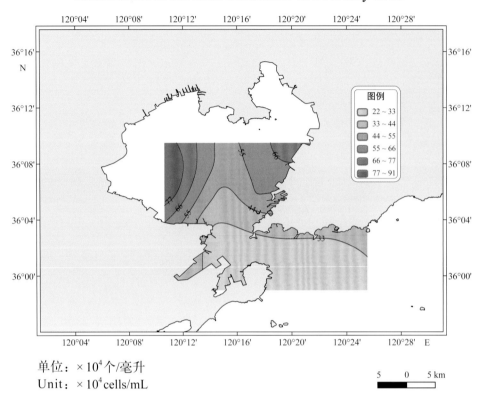

单位：×10⁴个/毫升
Unit：×10^{4}cells/mL

2010年5月水样含菌数分布
Distribution of bacterial abundance in May 2010

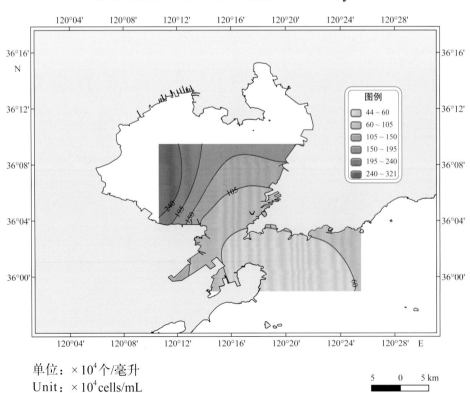

单位：×10⁴个/毫升
Unit：×10^{4}cells/mL

2010年8月水样含菌数分布
Distribution of bacterial abundance in August 2010

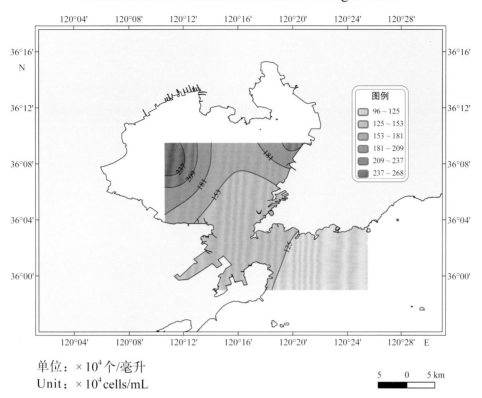

单位：×10⁴个/毫升
Unit：×10⁴ cells/mL

2010年11月水样含菌数分布
Distribution of bacterial abundance in November 2010

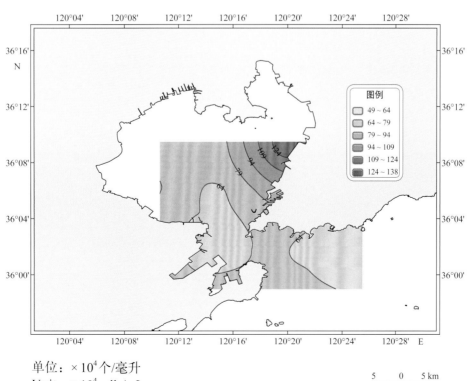

单位：×10⁴个/毫升
Unit：×10⁴ cells/mL

2011年2月水样含菌数分布
Distribution of bacterial abundance in February 2011

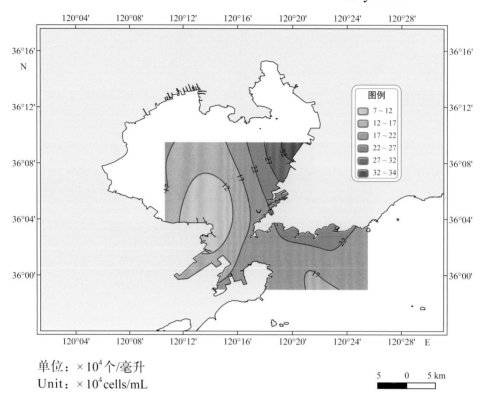

单位：×10⁴个/毫升
Unit：$\times 10^4$ cells/mL

2011年5月水样含菌数分布
Distribution of bacterial abundance in May 2011

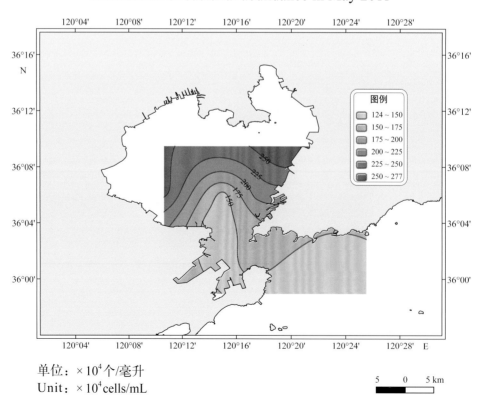

单位：×10⁴个/毫升
Unit：$\times 10^4$ cells/mL

2011年8月水样含菌数分布
Distribution of bacterial abundance in August 2011

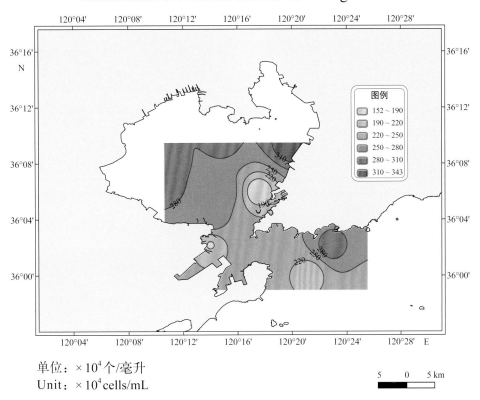

单位：×10⁴个/毫升
Unit：$\times 10^4$ cells/mL

2011年11月水样含菌数分布
Distribution of bacterial abundance in November 2011

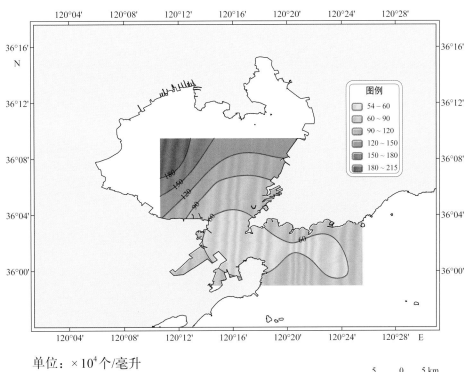

单位：×10⁴个/毫升
Unit：$\times 10^4$ cells/mL

2012年2月水样含菌数分布
Distribution of bacterial abundance in February 2012

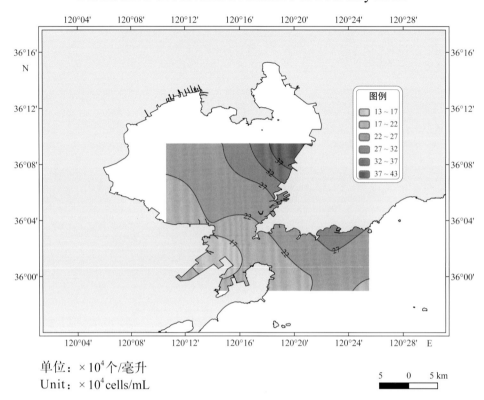

单位：$\times 10^4$ 个/毫升
Unit：$\times 10^4$ cells/mL

2012年5月水样含菌数分布
Distribution of bacterial abundance in May 2012

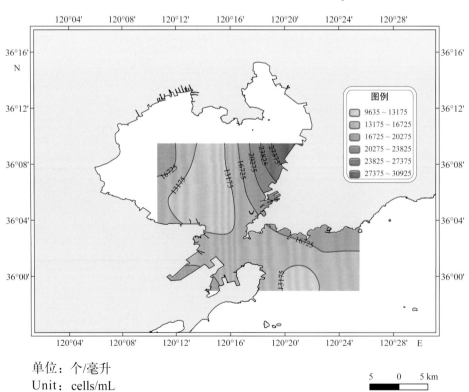

单位：个/毫升
Unit：cells/mL

2012年8月水样含菌数分布
Distribution of bacterial abundance in August 2012

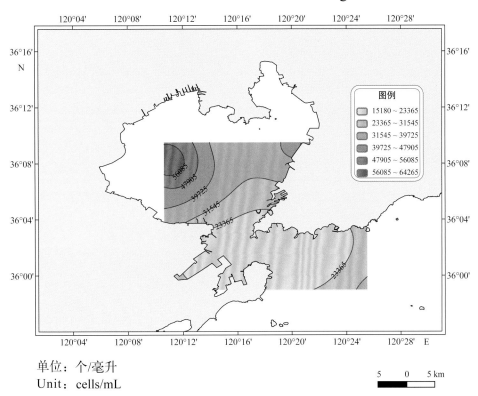

单位：个/毫升
Unit：cells/mL

2012年11月水样含菌数分布
Distribution of bacterial abundance in November 2012

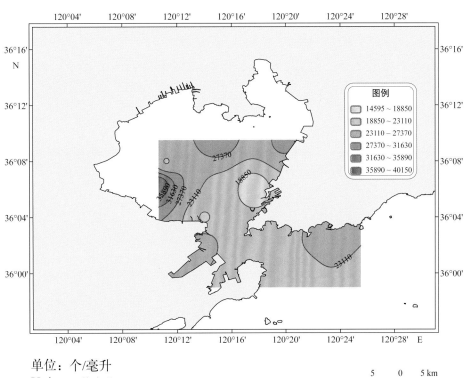

单位：个/毫升
Unit：cells/mL

2013年2月水样含菌数分布
Distribution of bacterial abundance in February 2013

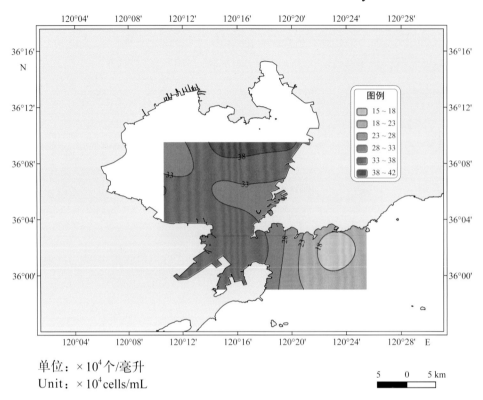

单位：$\times 10^4$ 个/毫升
Unit：$\times 10^4$ cells/mL

2013年5月水样含菌数分布
Distribution of bacterial abundance in May 2013

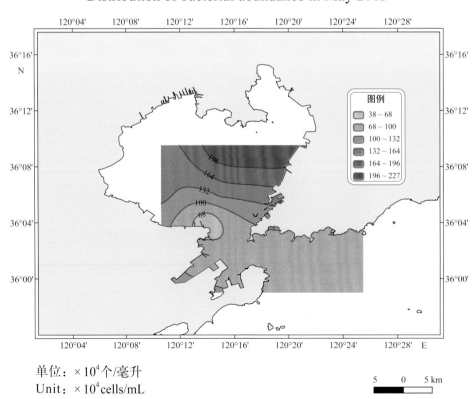

单位：$\times 10^4$ 个/毫升
Unit：$\times 10^4$ cells/mL

2013年8月水样含菌数分布
Distribution of bacterial abundance in August 2013

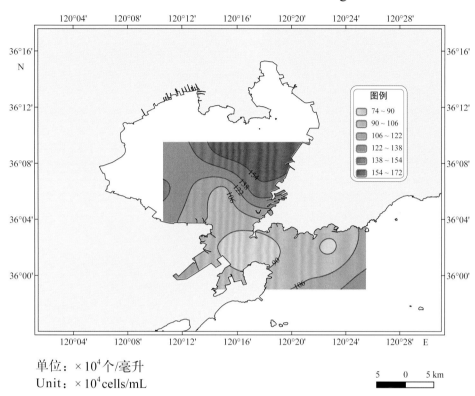

单位：×10⁴个/毫升
Unit：$\times 10^4$ cells/mL

2013年11月水样含菌数分布
Distribution of bacterial abundance in November 2013

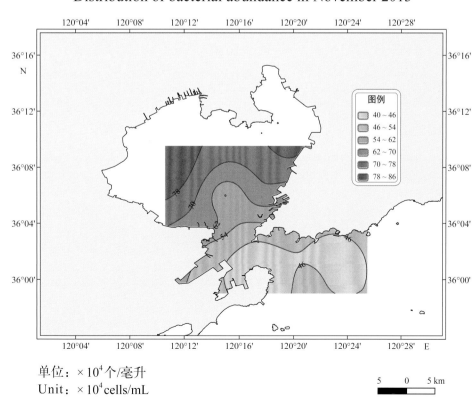

单位：×10⁴个/毫升
Unit：$\times 10^4$ cells/mL

2014年2月水样含菌数分布
Distribution of bacterial abundance in February 2014

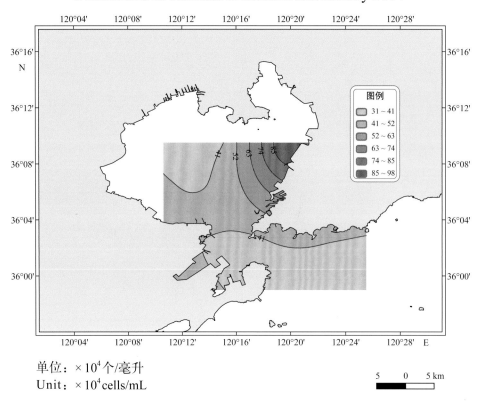

单位：×10⁴个/毫升
Unit: ×10⁴ cells/mL

2014年5月水样含菌数分布
Distribution of bacterial abundance in May 2014

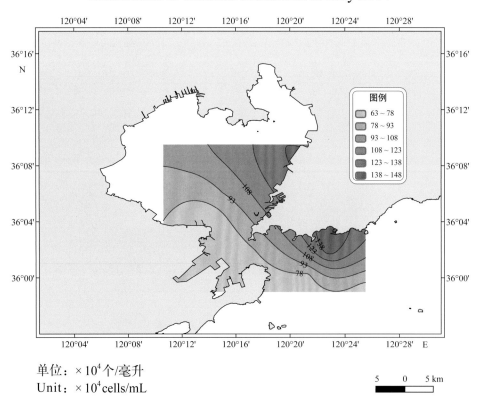

单位：×10⁴个/毫升
Unit: ×10⁴ cells/mL

2014年8月水样含菌数分布
Distribution of bacterial abundance in August 2014

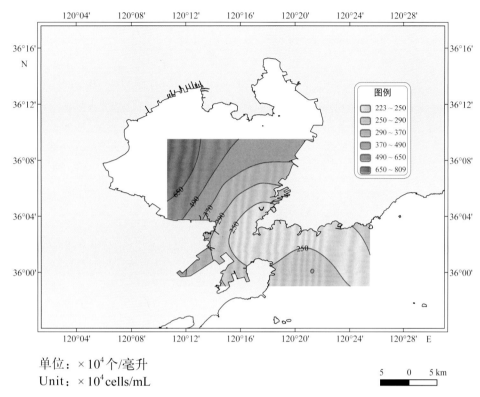

单位：$\times 10^4$ 个/毫升
Unit：$\times 10^4$ cells/mL

2014年11月水样含菌数分布
Distribution of bacterial abundance in November 2014

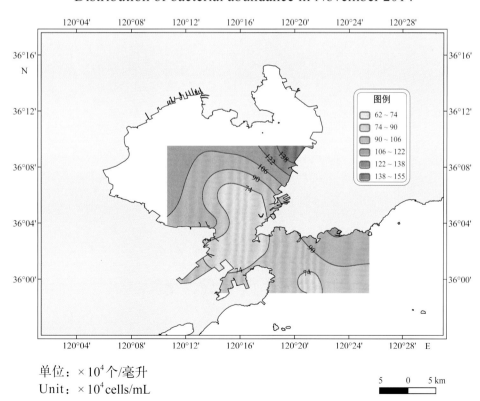

单位：$\times 10^4$ 个/毫升
Unit：$\times 10^4$ cells/mL

2015年2月水样含菌数分布
Distribution of bacterial abundance in February 2015

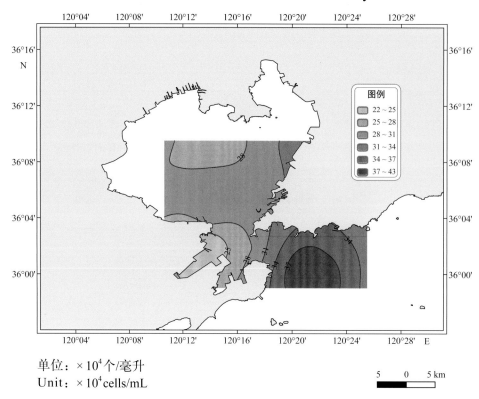

单位：$\times 10^4$ 个/毫升
Unit：$\times 10^4$ cells/mL

2015年5月水样含菌数分布
Distribution of bacterial abundance in May 2015

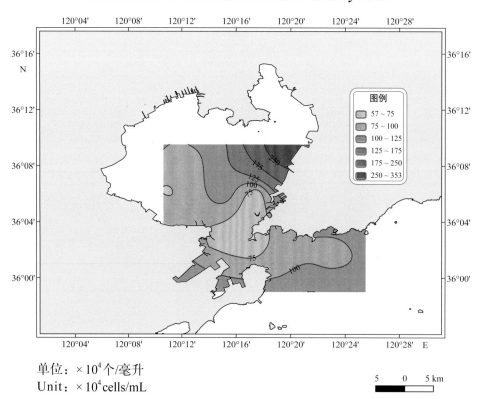

单位：$\times 10^4$ 个/毫升
Unit：$\times 10^4$ cells/mL

2015年8月水样含菌数分布
Distribution of bacterial abundance in August 2015

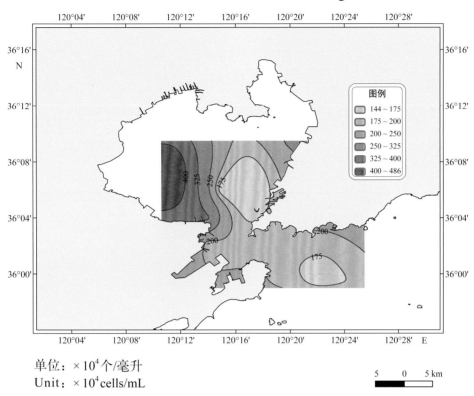

单位：×10^{4}个/毫升
Unit：×10^{4}cells/mL

2015年11月水样含菌数分布
Distribution of bacterial abundance in November 2015

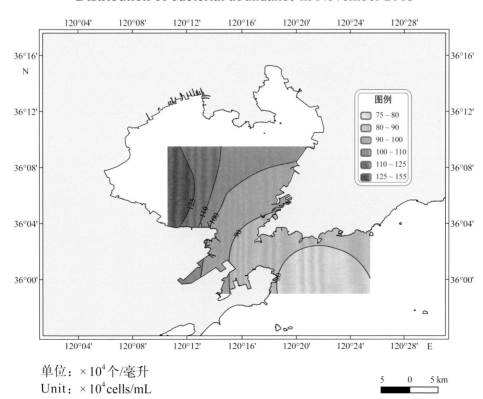

单位：×10^{4}个/毫升
Unit：×10^{4}cells/mL

7. 蓝细菌数 Cyanobacterial abundance

2010年2月蓝细菌数分布
Distribution of cyanobacterial abundance in February 2010

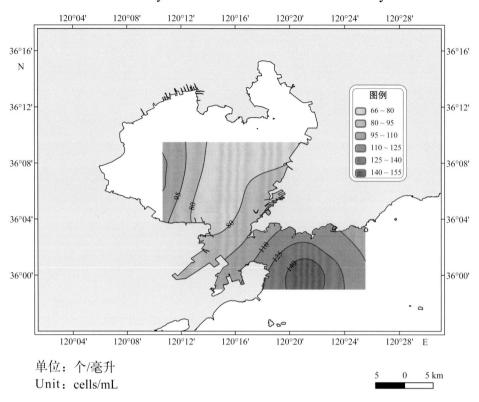

单位：个/毫升
Unit：cells/mL

2010年5月蓝细菌数分布
Distribution of cyanobacterial abundance in May 2010

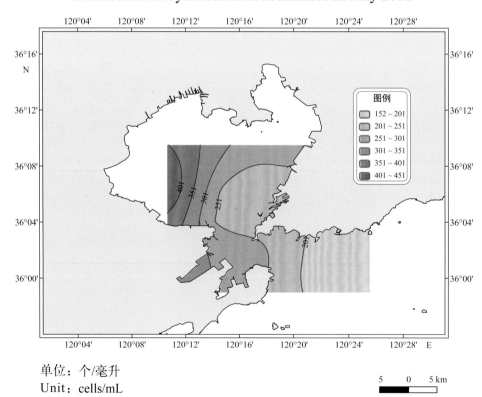

单位：个/毫升
Unit：cells/mL

2010年8月蓝细菌数分布
Distribution of cyanobacterial abundance in August 2010

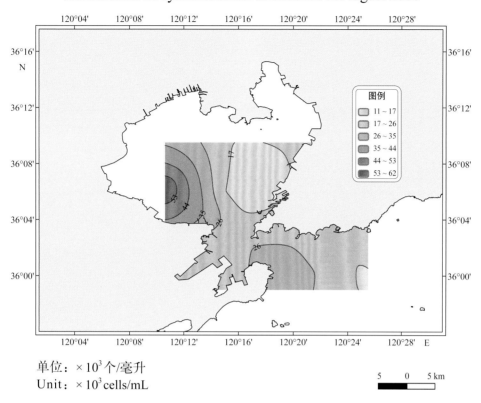

单位：$\times 10^3$ 个/毫升
Unit：$\times 10^3$ cells/mL

2010年11月蓝细菌数分布
Distribution of cyanobacterial abundance in November 2010

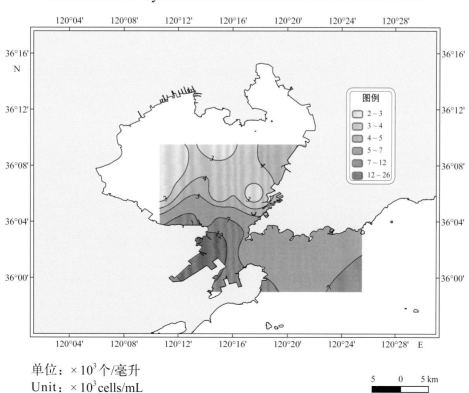

单位：$\times 10^3$ 个/毫升
Unit：$\times 10^3$ cells/mL

2011年2月蓝细菌数分布
Distribution of cyanobacterial abundance in February 2011

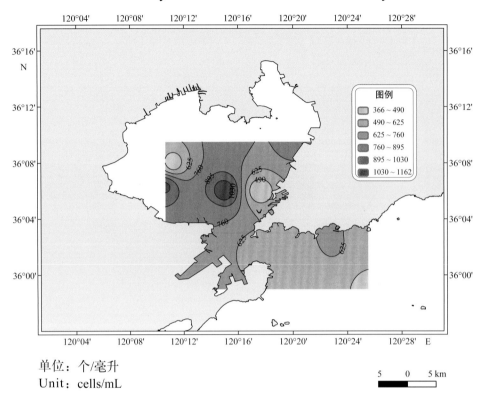

单位：个/毫升
Unit：cells/mL

2011年5月蓝细菌数分布
Distribution of cyanobacterial abundance in May 2011

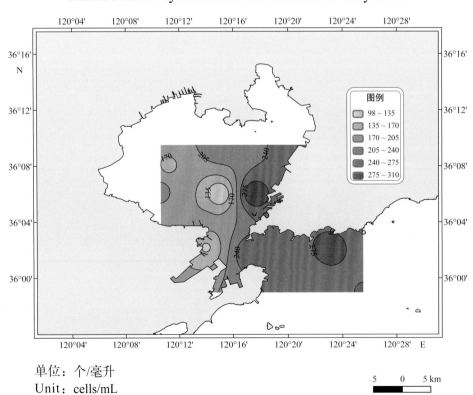

单位：个/毫升
Unit：cells/mL

2011年8月蓝细菌数分布
Distribution of cyanobacterial abundance in August 2011

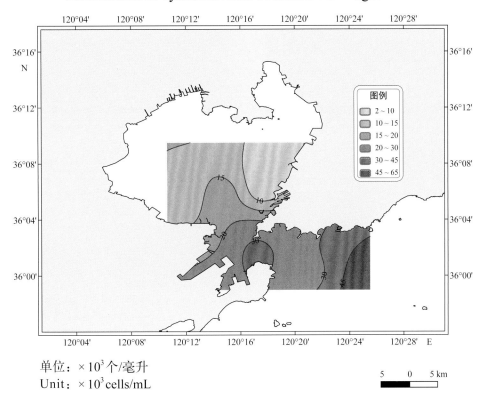

单位：×10³个/毫升
Unit：×10³ cells/mL

2011年11月蓝细菌数分布
Distribution of cyanobacterial abundance in November 2011

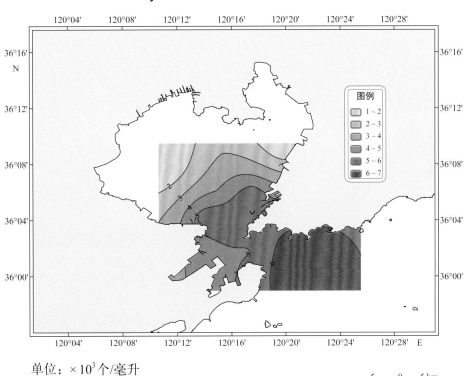

单位：×10³个/毫升
Unit：×10³ cells/mL

2012年2月蓝细菌数分布
Distribution of cyanobacterial abundance in February 2012

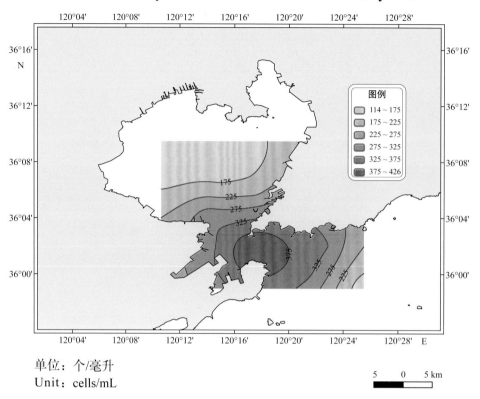

单位：个/毫升
Unit: cells/mL

2012年5月蓝细菌数分布
Distribution of cyanobacterial abundance in May 2012

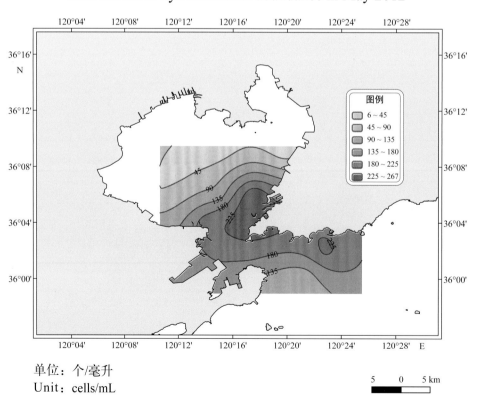

单位：个/毫升
Unit: cells/mL

2012年8月蓝细菌数分布
Distribution of cyanobacterial abundance in August 2012

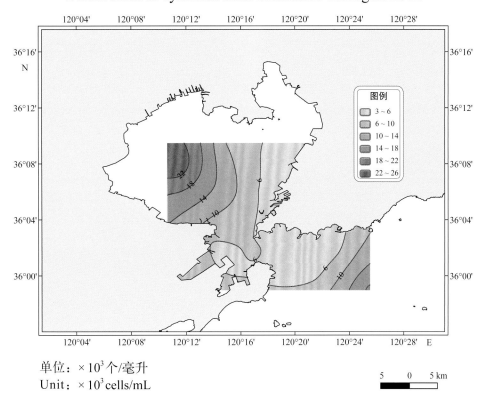

单位：$\times 10^3$ 个/毫升
Unit：$\times 10^3$ cells/mL

2012年11月蓝细菌数分布
Distribution of cyanobacterial abundance in November 2012

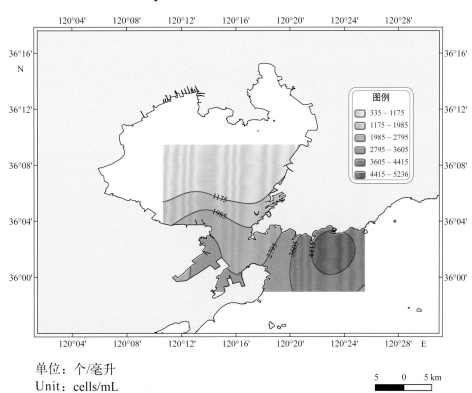

单位：个/毫升
Unit：cells/mL

2013年2月蓝细菌数分布
Distribution of cyanobacterial abundance in February 2013

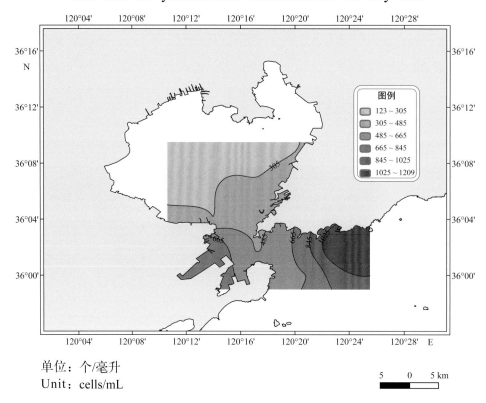

单位：个/毫升
Unit：cells/mL

2013年5月蓝细菌数分布
Distribution of cyanobacterial abundance in May 2013

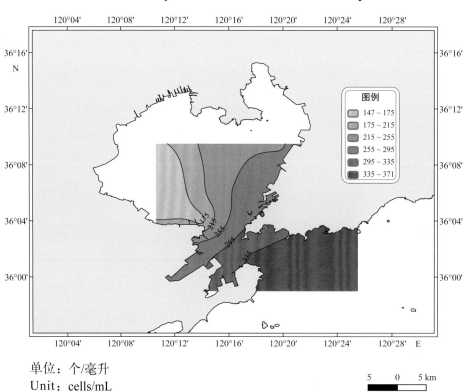

单位：个/毫升
Unit：cells/mL

2013年8月蓝细菌数分布
Distribution of cyanobacterial abundance in August 2013

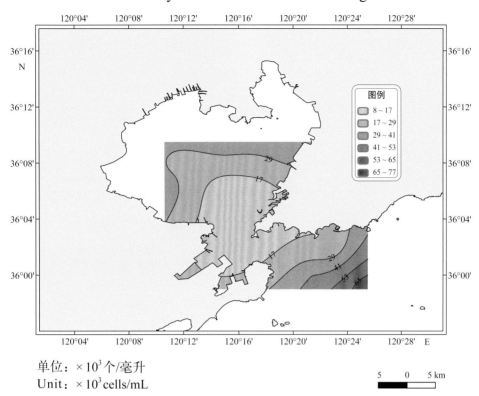

单位：×10³个/毫升
Unit：×10³cells/mL

2013年11月蓝细菌数分布
Distribution of cyanobacterial abundance in November 2013

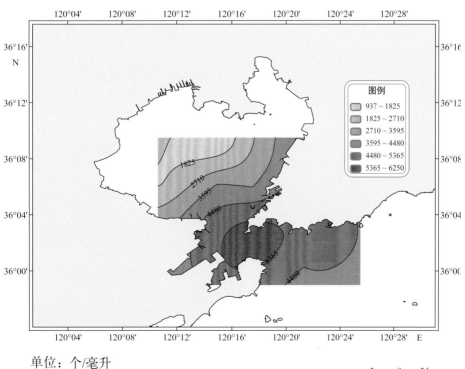

单位：个/毫升
Unit：cells/mL

2014年2月蓝细菌数分布
Distribution of cyanobacterial abundance in February 2014

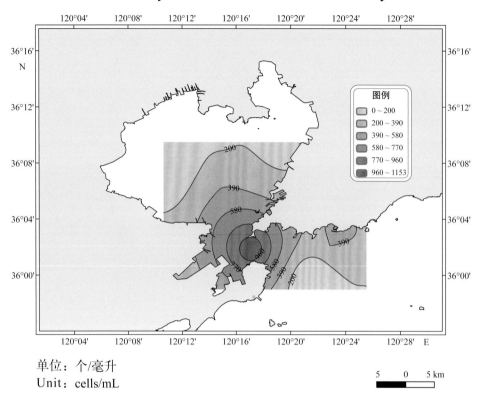

单位：个/毫升
Unit：cells/mL

2014年5月蓝细菌数分布
Distribution of cyanobacterial abundance in May 2014

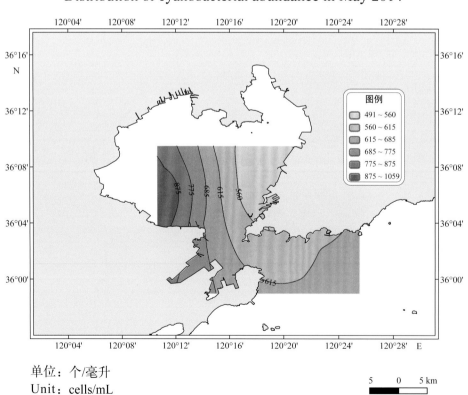

单位：个/毫升
Unit：cells/mL

2014年8月蓝细菌数分布
Distribution of cyanobacterial abundance in August 2014

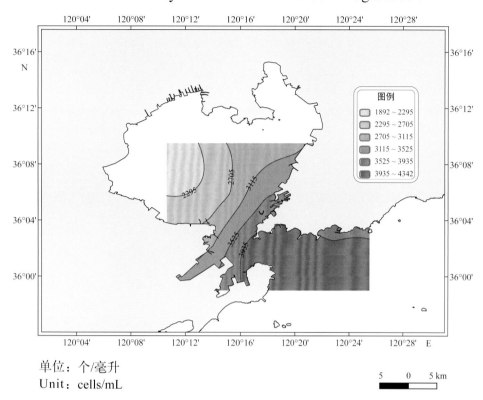

单位：个/毫升
Unit: cells/mL

2014年11月蓝细菌数分布
Distribution of cyanobacterial abundance in November 2014

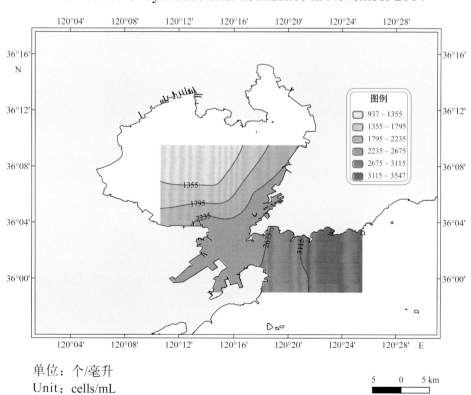

单位：个/毫升
Unit: cells/mL

2015年2月蓝细菌数分布
Distribution of cyanobacterial abundance in February 2015

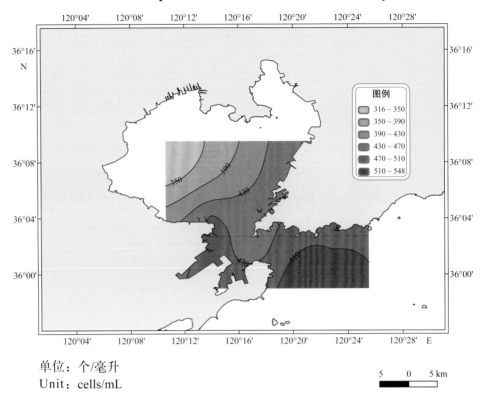

单位：个/毫升
Unit：cells/mL

2015年5月蓝细菌数分布
Distribution of cyanobacterial abundance in May 2015

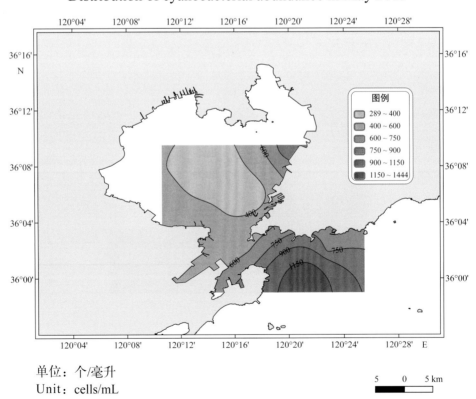

单位：个/毫升
Unit：cells/mL

2015年8月蓝细菌数分布
Distribution of cyanobacterial abundance in August 2015

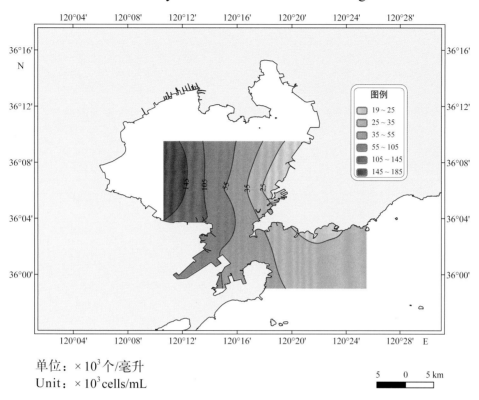

单位：×10³个/毫升
Unit：×10³ cells/mL

2015年11月蓝细菌数分布
Distribution of cyanobacterial abundance in November 2015

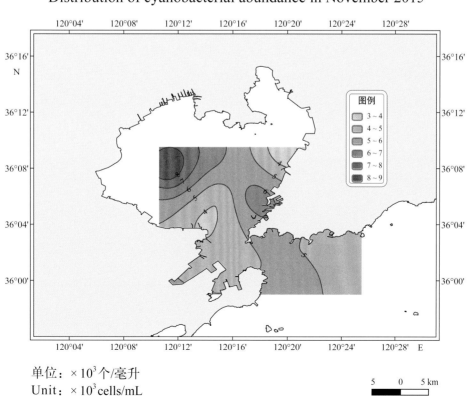

单位：×10³个/毫升
Unit：×10³ cells/mL

8. 大肠杆菌丰度 Coliform bacteria abundance

2010年2月大肠杆菌丰度分布
Distribution of coliform bacteria abundance in February 2010

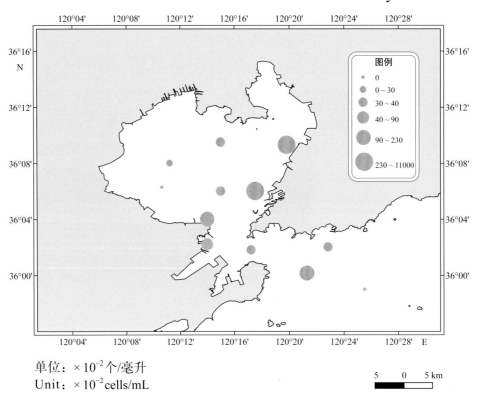

单位：×10^{-2} 个/毫升
Unit：×10^{-2} cells/mL

2010年5月大肠杆菌丰度分布
Distribution of coliform bacteria abundance in May 2010

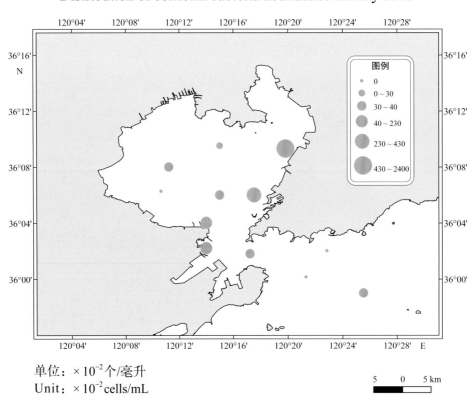

单位：×10^{-2} 个/毫升
Unit：×10^{-2} cells/mL

2010年8月大肠杆菌丰度分布
Distribution of coliform bacteria abundance in August 2010

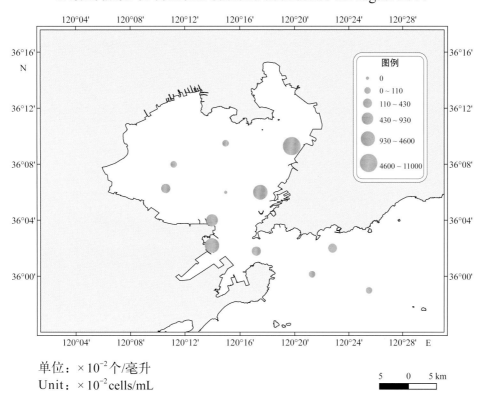

单位：×10^{-2}个/毫升
Unit：×10^{-2}cells/mL

2010年11月大肠杆菌丰度分布
Distribution of coliform bacteria abundance in November 2010

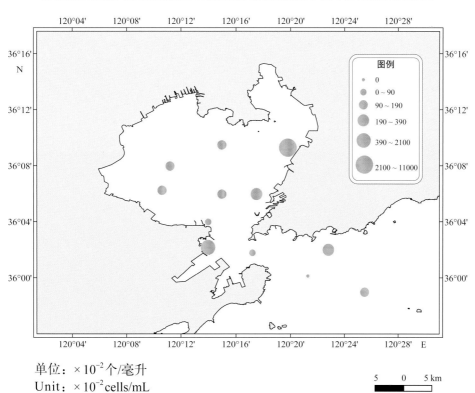

单位：×10^{-2}个/毫升
Unit：×10^{-2}cells/mL

2011年2月大肠杆菌丰度分布
Distribution of coliform bacteria abundance in February 2011

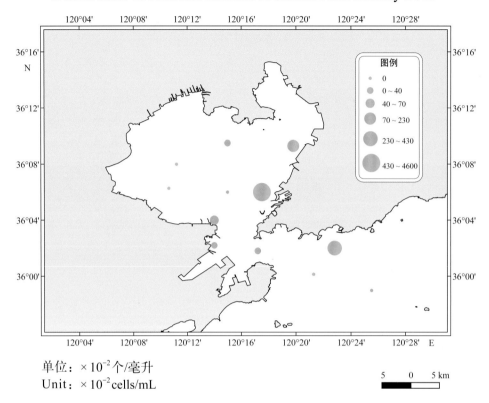

单位：×10^{-2} 个/毫升
Unit：×10^{-2} cells/mL

2011年5月大肠杆菌丰度分布
Distribution of coliform bacteria abundance in May 2011

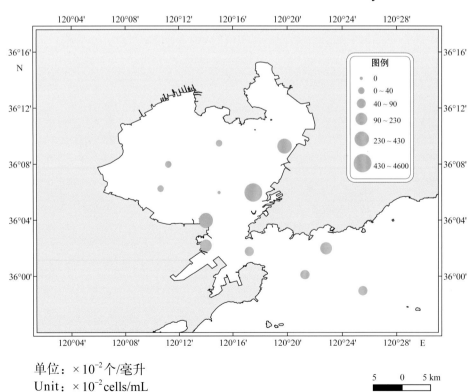

单位：×10^{-2} 个/毫升
Unit：×10^{-2} cells/mL

2011年8月大肠杆菌丰度分布
Distribution of coliform bacteria abundance in August 2011

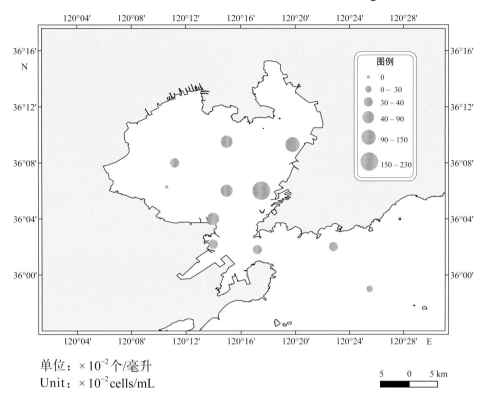

单位：×10^{-2}个/毫升
Unit：×10^{-2}cells/mL

2011年11月大肠杆菌丰度分布
Distribution of coliform bacteria abundance in November 2011

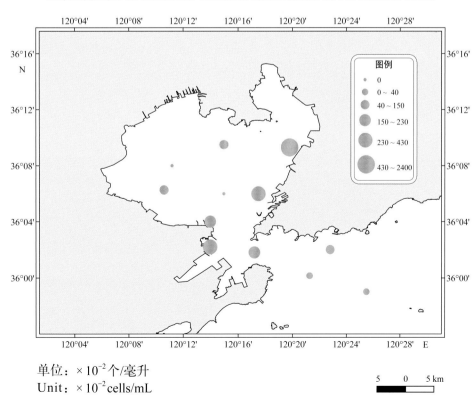

单位：×10^{-2}个/毫升
Unit：×10^{-2}cells/mL

2012年2月大肠杆菌丰度分布
Distribution of coliform bacteria abundance in February 2012

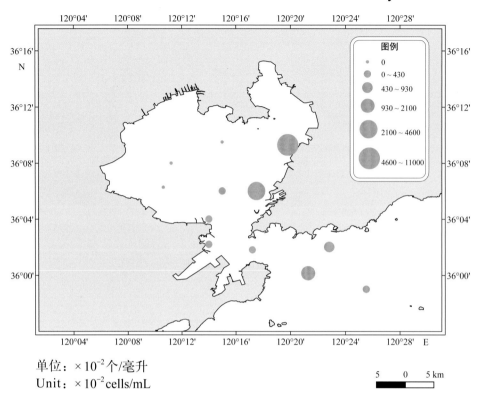

单位：×10^{-2}个/毫升
Unit：×10^{-2} cells/mL

2012年5月大肠杆菌丰度分布
Distribution of coliform bacteria abundance in May 2012

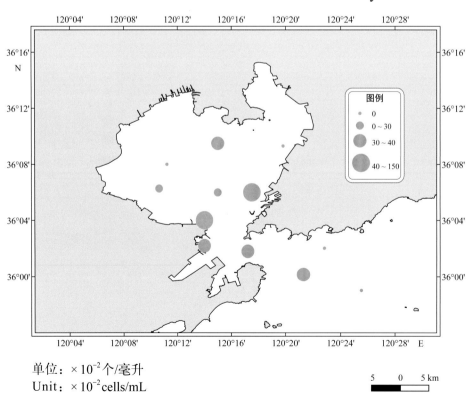

单位：×10^{-2}个/毫升
Unit：×10^{-2} cells/mL

2012年8月大肠杆菌丰度分布
Distribution of coliform bacteria abundance in August 2012

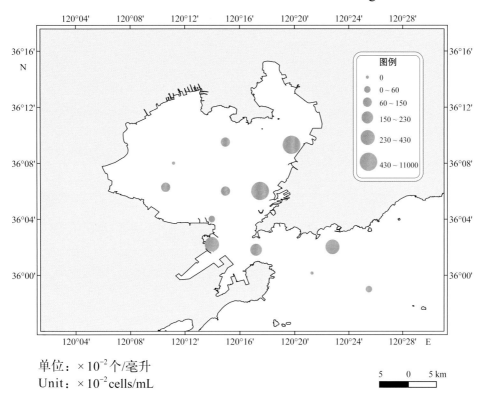

单位：×10⁻²个/毫升
Unit：×10⁻²cells/mL

2012年11月大肠杆菌丰度分布
Distribution of coliform bacteria abundance in November 2012

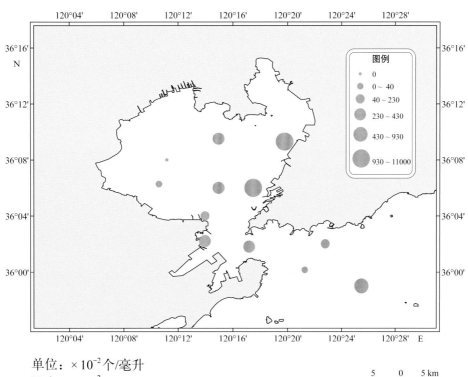

单位：×10⁻²个/毫升
Unit：×10⁻²cells/mL

2013年2月大肠杆菌丰度分布
Distribution of coliform bacteria abundance in February 2013

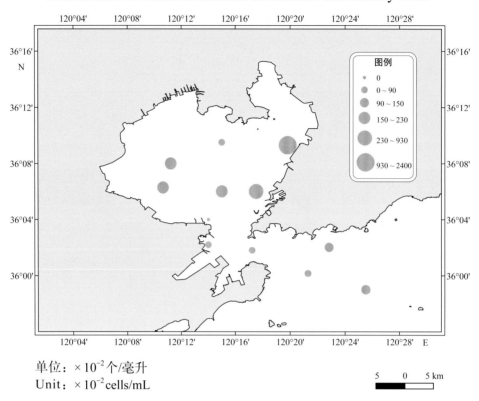

单位：×10⁻² 个/毫升
Unit: $\times 10^{-2}$ cells/mL

2013年5月大肠杆菌丰度分布
Distribution of coliform bacteria abundance in May 2013

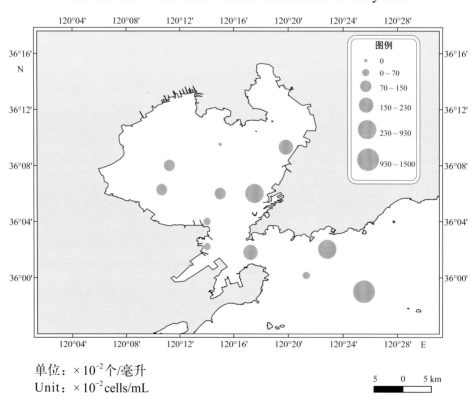

单位：×10⁻² 个/毫升
Unit: $\times 10^{-2}$ cells/mL

2013年8月大肠杆菌丰度分布
Distribution of coliform bacteria abundance in August 2013

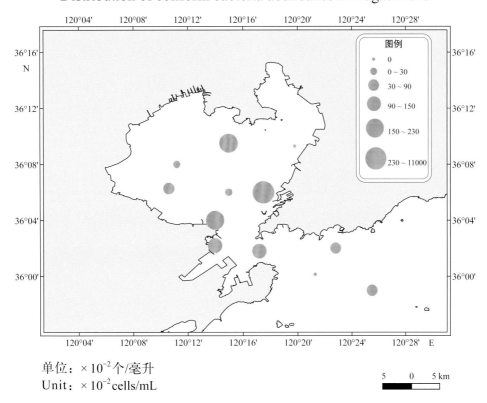

单位：$\times 10^{-2}$个/毫升
Unit：$\times 10^{-2}$ cells/mL

2013年11月大肠杆菌丰度分布
Distribution of coliform bacteria abundance in November 2013

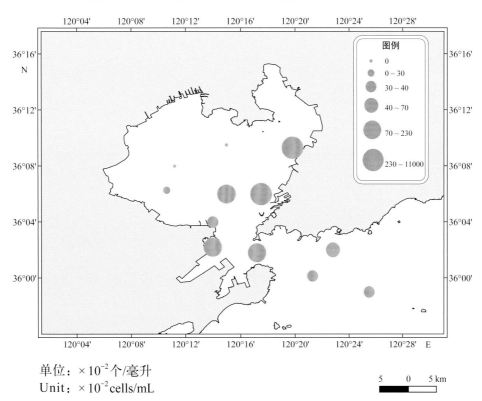

单位：$\times 10^{-2}$个/毫升
Unit：$\times 10^{-2}$ cells/mL

2014年2月大肠杆菌丰度分布
Distribution of coliform bacteria abundance in February 2014

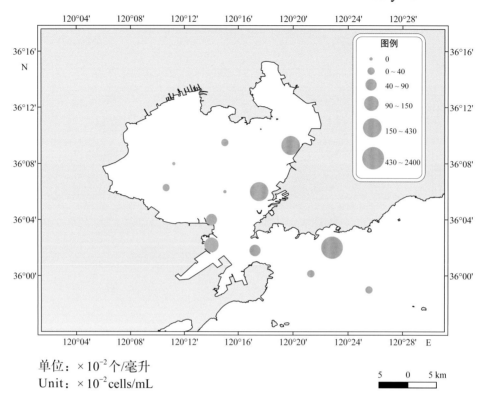

单位：×10^{-2}个/毫升
Unit：×10^{-2}cells/mL

2014年5月大肠杆菌丰度分布
Distribution of coliform bacteria abundance in May 2014

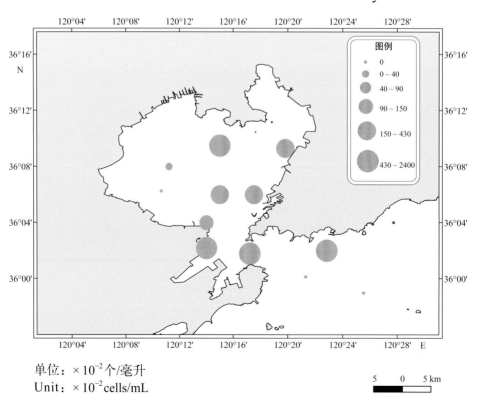

单位：×10^{-2}个/毫升
Unit：×10^{-2}cells/mL

2014年8月大肠杆菌丰度分布
Distribution of coliform bacteria abundance in August 2014

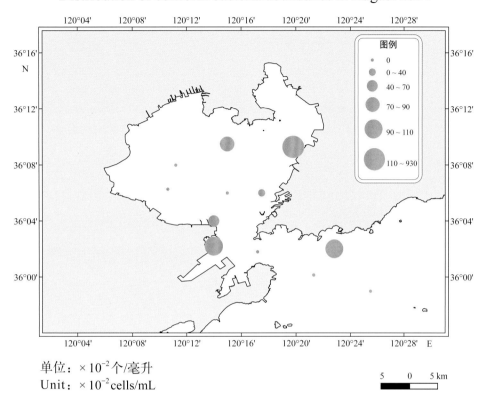

单位：$\times 10^{-2}$ 个/毫升
Unit：$\times 10^{-2}$ cells/mL

2014年11月大肠杆菌丰度分布
Distribution of coliform bacteria abundance in November 2014

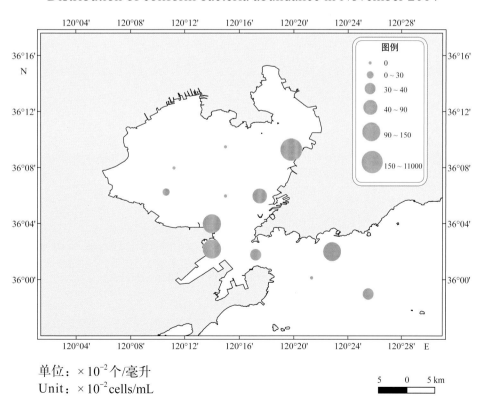

单位：$\times 10^{-2}$ 个/毫升
Unit：$\times 10^{-2}$ cells/mL

2015年2月大肠杆菌丰度分布
Distribution of coliform bacteria abundance in February 2015

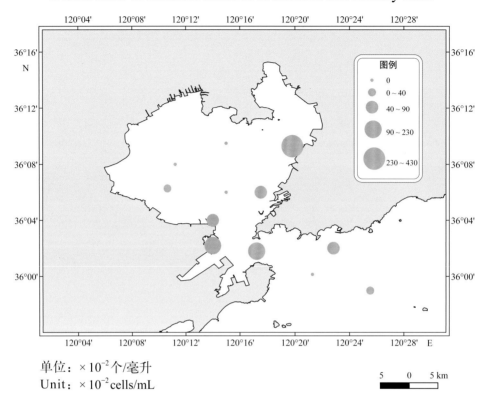

单位：$\times 10^{-2}$ 个/毫升
Unit：$\times 10^{-2}$ cells/mL

2015年5月大肠杆菌丰度分布
Distribution of coliform bacteria abundance in May 2015

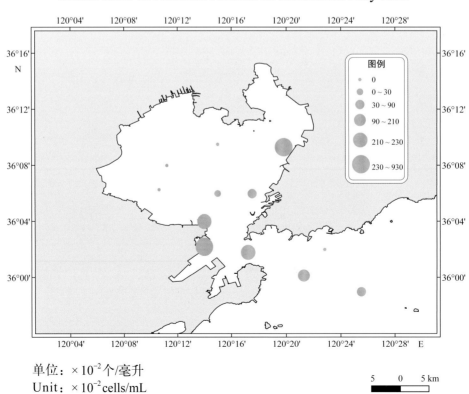

单位：$\times 10^{-2}$ 个/毫升
Unit：$\times 10^{-2}$ cells/mL

2015年8月大肠杆菌丰度分布
Distribution of coliform bacteria abundance in August 2015

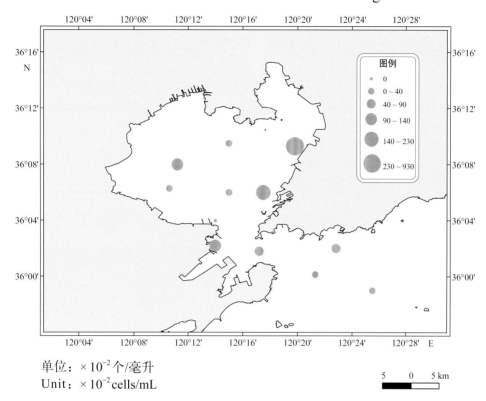

单位：$\times 10^{-2}$ 个/毫升
Unit：$\times 10^{-2}$ cells/mL

2015年11月大肠杆菌丰度分布
Distribution of coliform bacteria abundance in November 2015

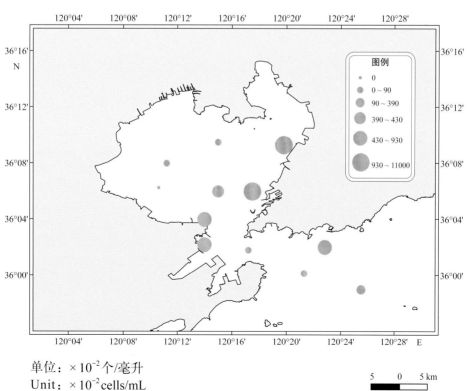

单位：$\times 10^{-2}$ 个/毫升
Unit：$\times 10^{-2}$ cells/mL

9. 底栖生物总生物量 Total biomass of benthos

2010年2月底栖生物总生物量分布
Distribution of total benthic biomass in February 2010

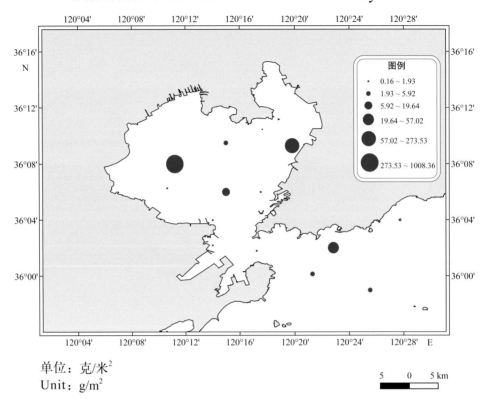

单位：克/米²
Unit：g/m²

2010年5月底栖生物总生物量分布
Distribution of total benthic biomass in May 2010

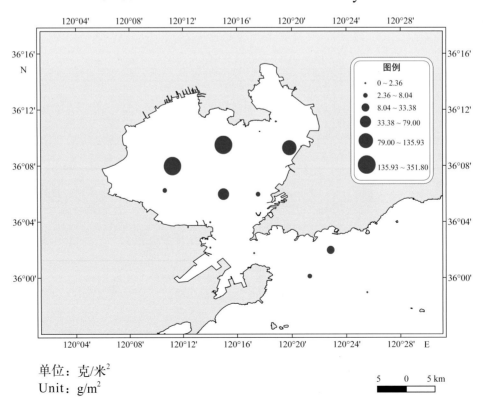

单位：克/米²
Unit：g/m²

2010年8月底栖生物总生物量分布
Distribution of total benthic biomass in August 2010

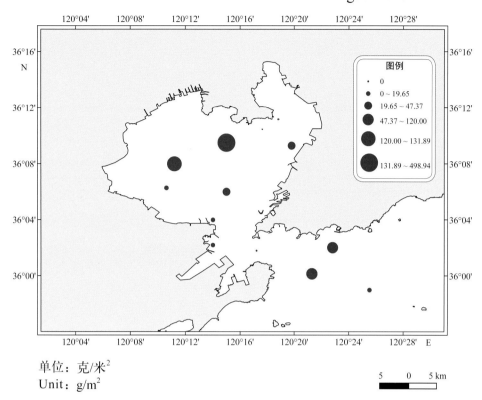

单位：克/米²
Unit：g/m²

2010年11月底栖生物总生物量分布
Distribution of total benthic biomass in November 2010

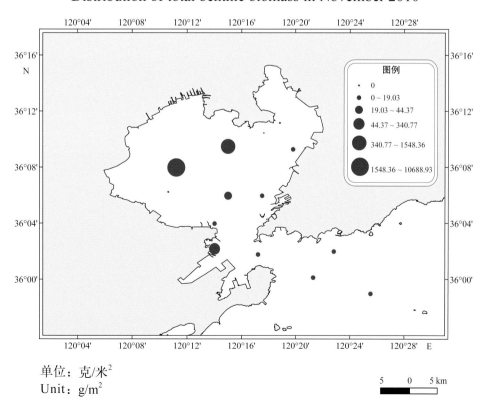

单位：克/米²
Unit：g/m²

2011年2月底栖生物总生物量分布
Distribution of total benthic biomass in February 2011

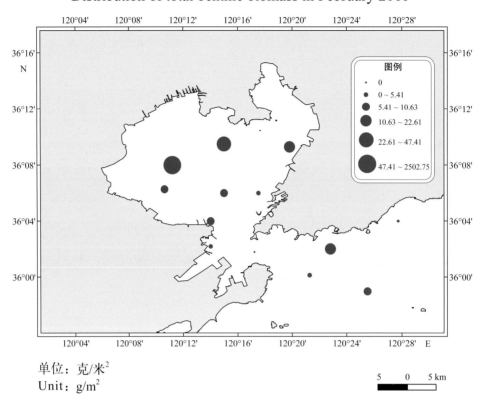

单位：克/米²
Unit：g/m²

2011年5月底栖生物总生物量分布
Distribution of total benthic biomass in May 2011

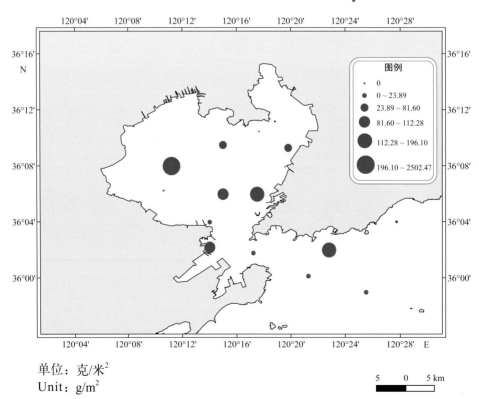

单位：克/米²
Unit：g/m²

2011年8月底栖生物总生物量分布
Distribution of total benthic biomass in August 2011

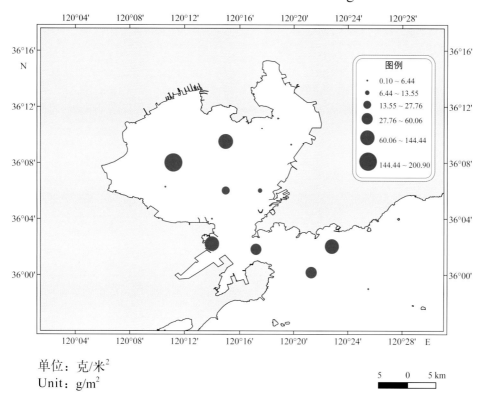

单位：克/米^2
Unit：g/m^2

2011年11月底栖生物总生物量分布
Distribution of total benthic biomass in November 2011

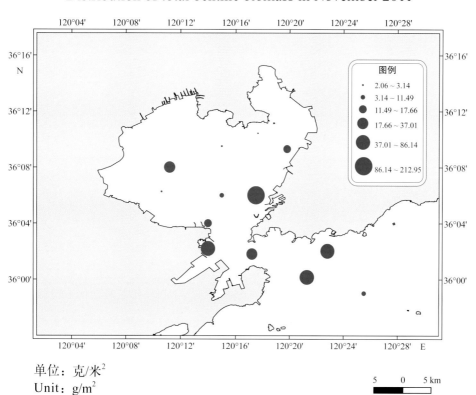

单位：克/米^2
Unit：g/m^2

2012年2月底栖生物总生物量分布
Distribution of total benthic biomass in February 2012

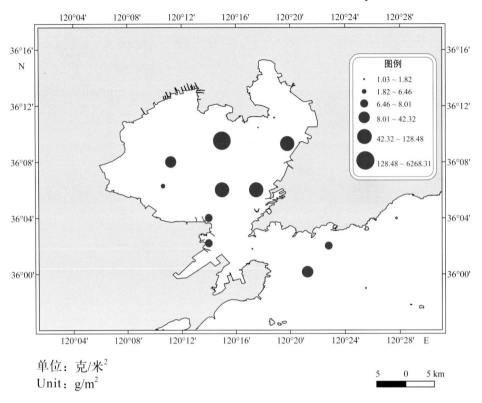

单位：克/米²
Unit：g/m²

2012年5月底栖生物总生物量分布
Distribution of total benthic biomass in May 2012

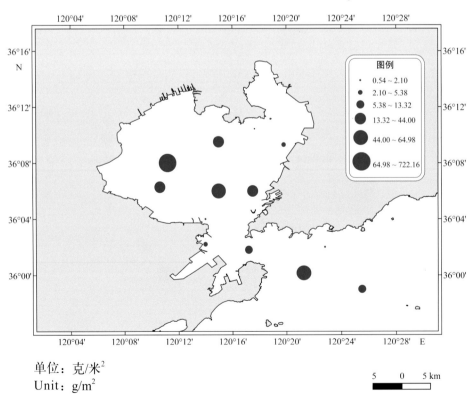

单位：克/米²
Unit：g/m²

2012年8月底栖生物总生物量分布
Distribution of total benthic biomass in August 2012

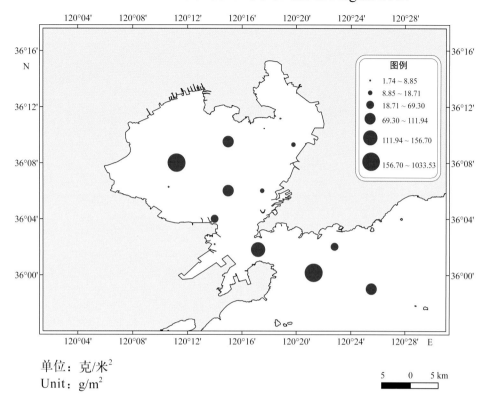

单位：克/米²
Unit：g/m²

2012年11月底栖生物总生物量分布
Distribution of total benthic biomass in November 2012

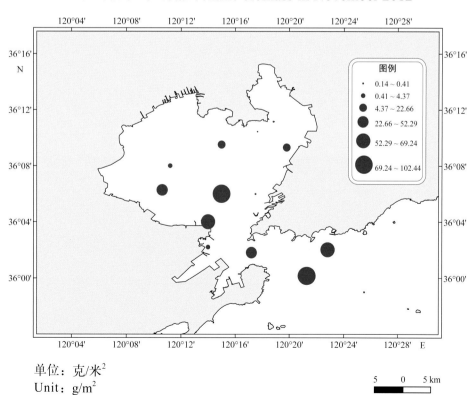

单位：克/米²
Unit：g/m²

2013年2月底栖生物总生物量分布
Distribution of total benthic biomass in February 2013

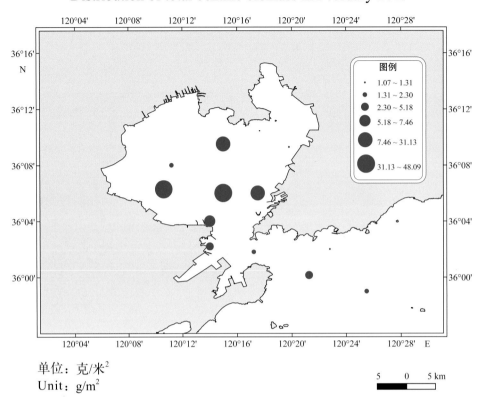

单位：克/米²
Unit：g/m²

2013年5月底栖生物总生物量分布
Distribution of total benthic biomass in May 2013

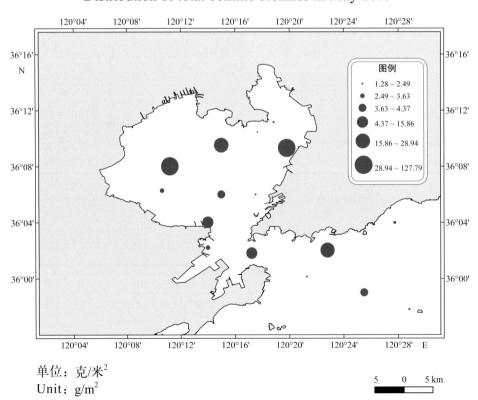

单位：克/米²
Unit：g/m²

2013年8月底栖生物总生物量分布
Distribution of total benthic biomass in August 2013

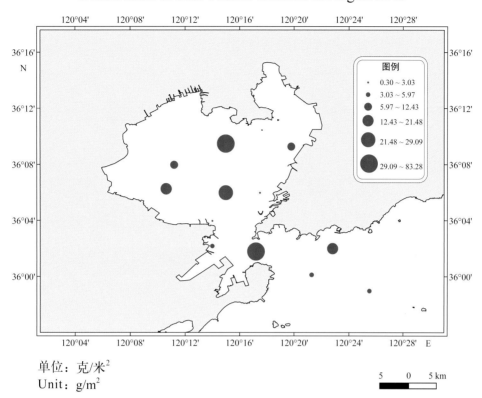

单位：克/米²
Unit：g/m²

2013年11月底栖生物总生物量分布
Distribution of total benthic biomass in November 2013

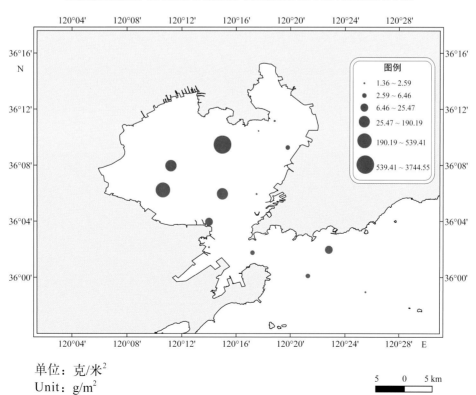

单位：克/米²
Unit：g/m²

2014年2月底栖生物总生物量分布
Distribution of total benthic biomass in February 2014

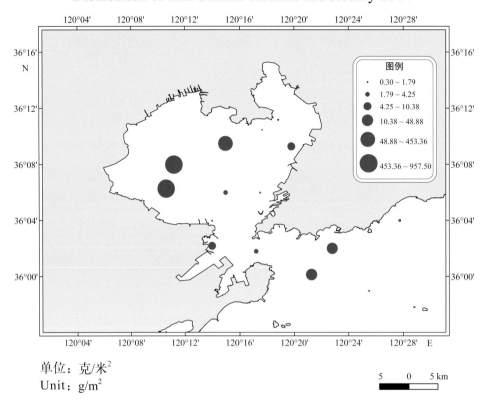

单位：克/米²
Unit：g/m²

2014年5月底栖生物总生物量分布
Distribution of total benthic biomass in May 2014

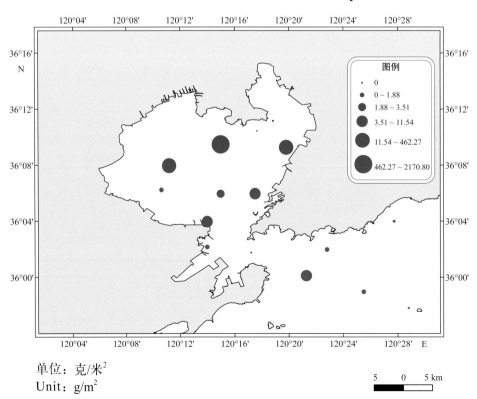

单位：克/米²
Unit：g/m²

2014年8月底栖生物总生物量分布
Distribution of total benthic biomass in August 2014

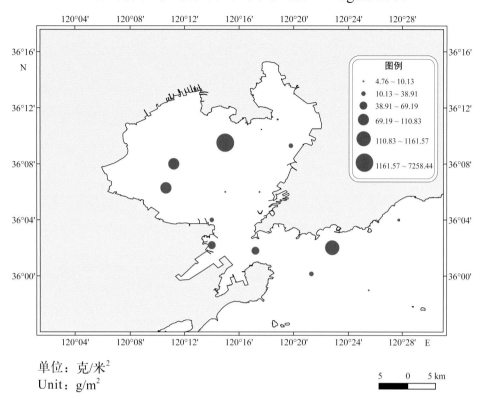

单位：克/米²
Unit：g/m²

2014年11月底栖生物总生物量分布
Distribution of total benthic biomass in November 2014

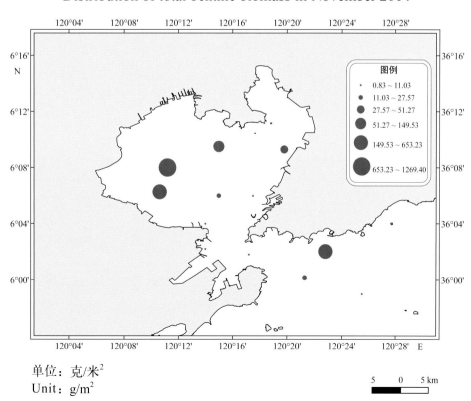

单位：克/米²
Unit：g/m²

2015年2月底栖生物总生物量分布
Distribution of total benthic biomass in February 2015

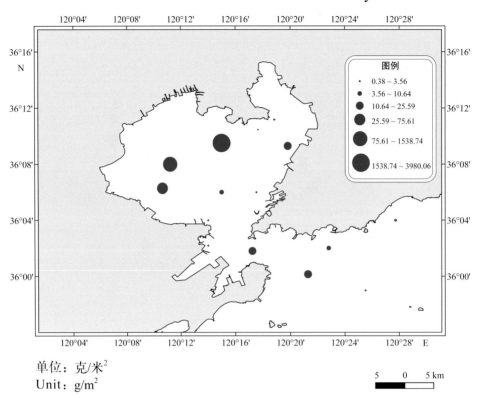

单位：克/米²
Unit：g/m²

2015年5月底栖生物总生物量分布
Distribution of total benthic biomass in May 2015

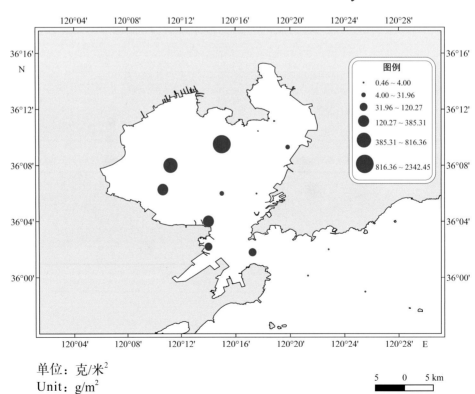

单位：克/米²
Unit：g/m²

2015年8月底栖生物总生物量分布
Distribution of total benthic biomass in August 2015

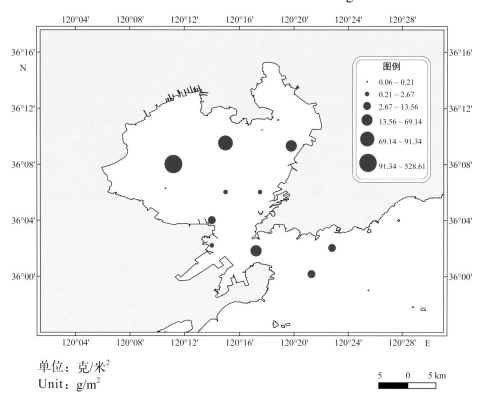

单位：克/米^2
Unit：g/m^2

2015年11月底栖生物总生物量分布
Distribution of total benthic biomass in November 2015

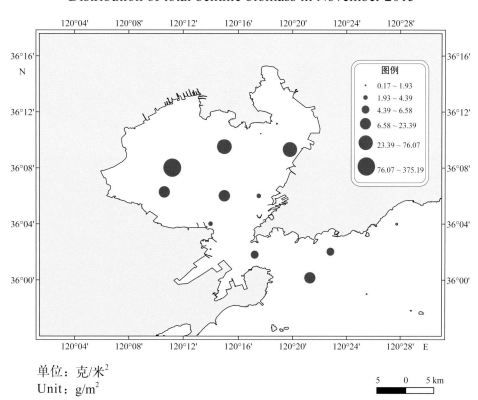

单位：克/米^2
Unit：g/m^2

10. 底栖生物总密度 Total density of benthos

2010年2月底栖生物总密度分布
Distribution of total benthic density in February 2010

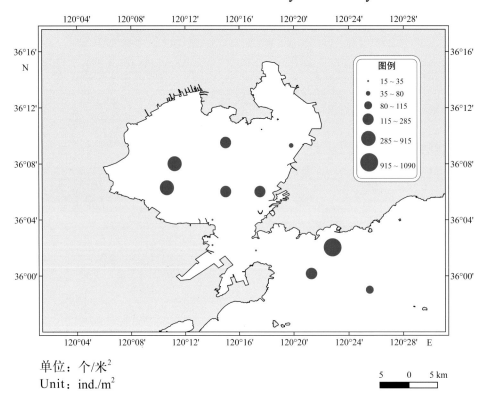

单位：个/米²
Unit: ind./m²

2010年5月底栖生物总密度分布
Distribution of total benthic density in May 2010

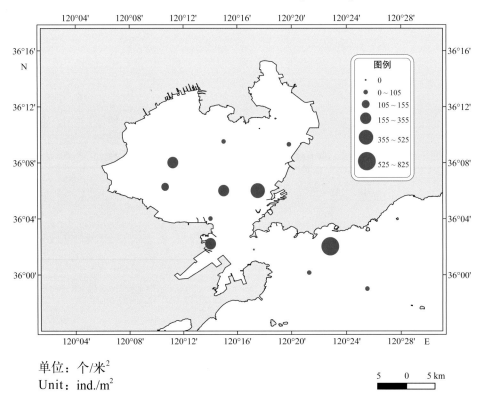

单位：个/米²
Unit: ind./m²

2010年8月底栖生物总密度分布
Distribution of total benthic density in August 2010

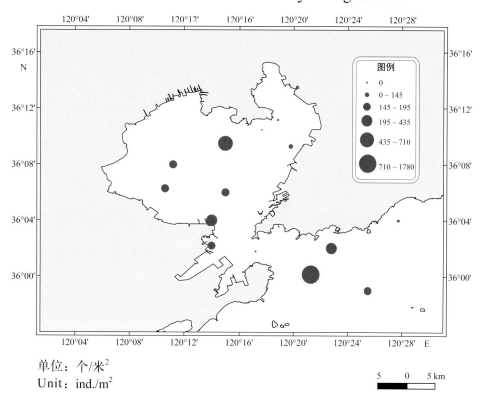

单位：个/米²
Unit：ind./m²

2010年11月底栖生物总密度分布
Distribution of total benthic density in November 2010

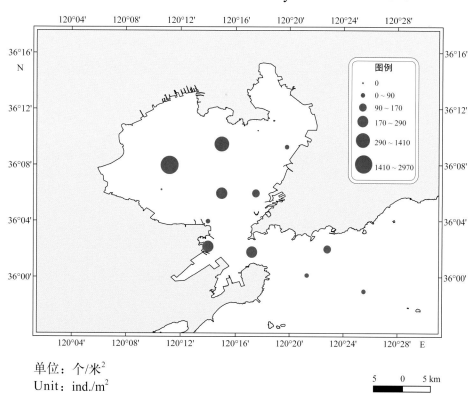

单位：个/米²
Unit：ind./m²

2011年2月底栖生物总密度分布
Distribution of total benthic density in February 2011

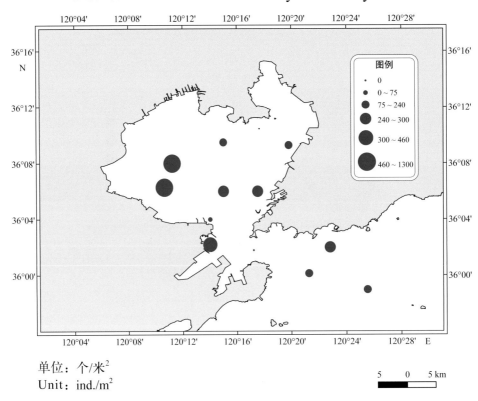

单位：个/米²
Unit：ind./m²

2011年5月底栖生物总密度分布
Distribution of total benthic density in May 2011

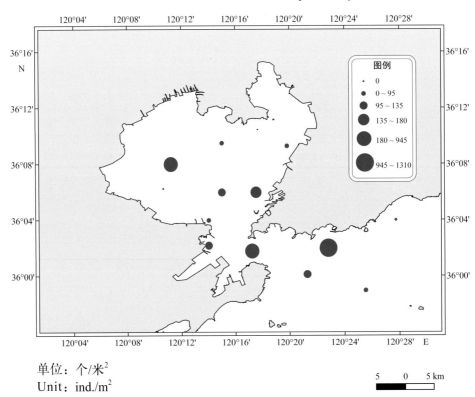

单位：个/米²
Unit：ind./m²

2011年8月底栖生物总密度分布
Distribution of total benthic density in August 2011

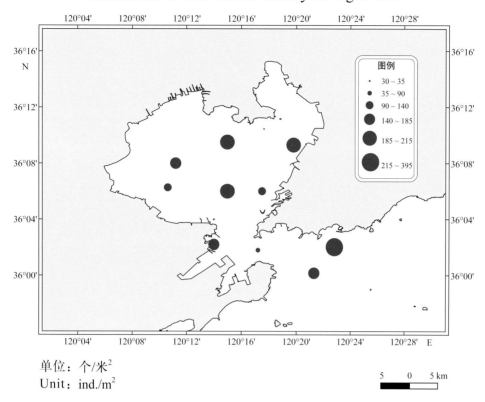

单位：个/米²
Unit：ind./m²

2011年11月底栖生物总密度分布
Distribution of total benthic density in November 2011

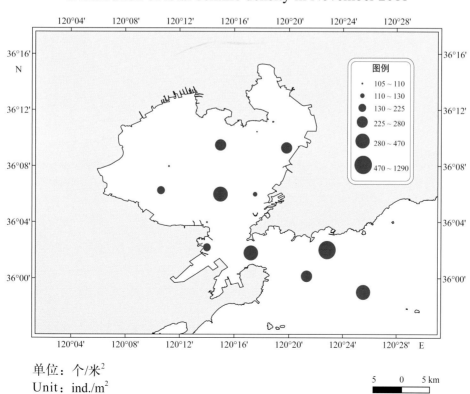

单位：个/米²
Unit：ind./m²

2012年2月底栖生物总密度分布
Distribution of total benthic density in February 2012

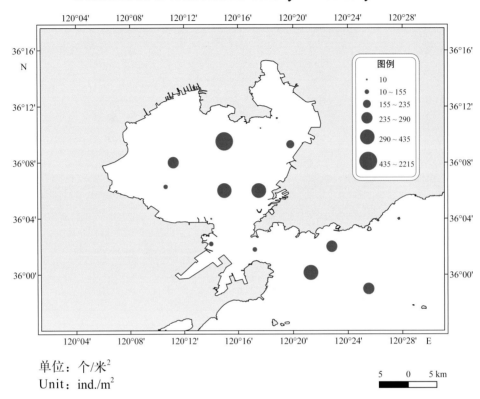

单位：个/米²
Unit：ind./m²

2012年5月底栖生物总密度分布
Distribution of total benthic density in May 2012

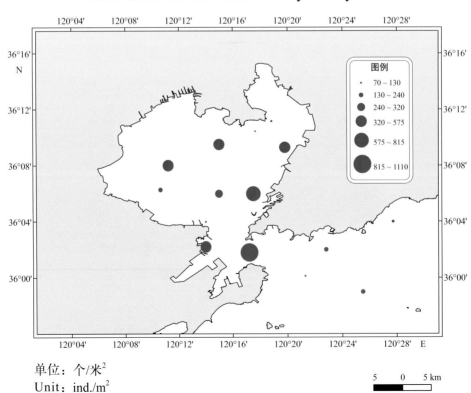

单位：个/米²
Unit：ind./m²

2012年8月底栖生物总密度分布
Distribution of total benthic density in August 2012

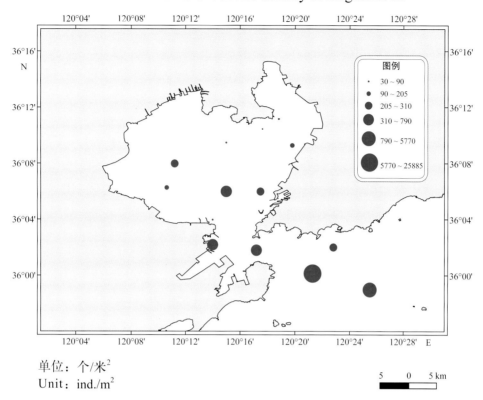

单位：个/米²
Unit：ind./m²

2012年11月底栖生物总密度分布
Distribution of total benthic density in November 2012

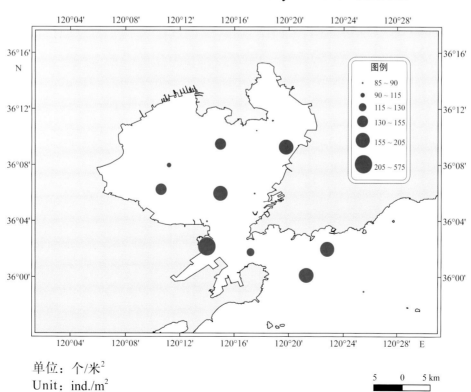

单位：个/米²
Unit：ind./m²

2013年2月底栖生物总密度分布
Distribution of total benthic density in February 2013

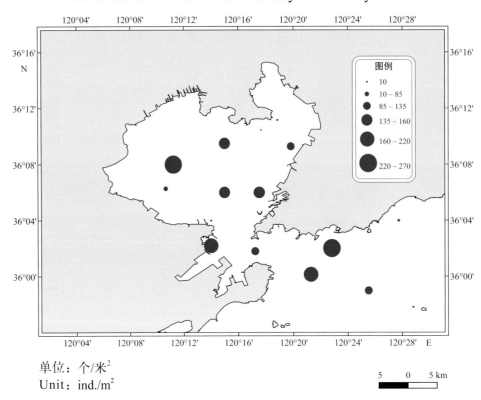

单位：个/米²
Unit：ind./m²

2013年5月底栖生物总密度分布
Distribution of total benthic density in May 2013

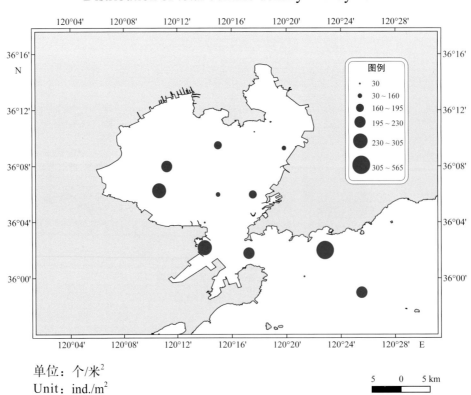

单位：个/米²
Unit：ind./m²

2013年8月底栖生物总密度分布
Distribution of total benthic density in August 2013

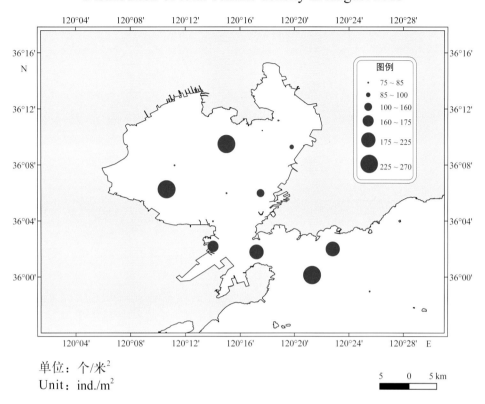

单位：个/米²
Unit：ind./m²

2013年11月底栖生物总密度分布
Distribution of total benthic density in November 2013

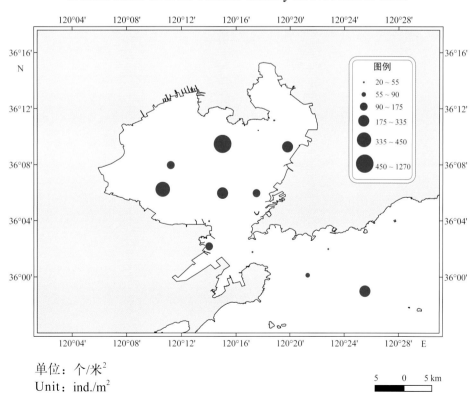

单位：个/米²
Unit：ind./m²

2014年2月底栖生物总密度分布
Distribution of total benthic density in February 2014

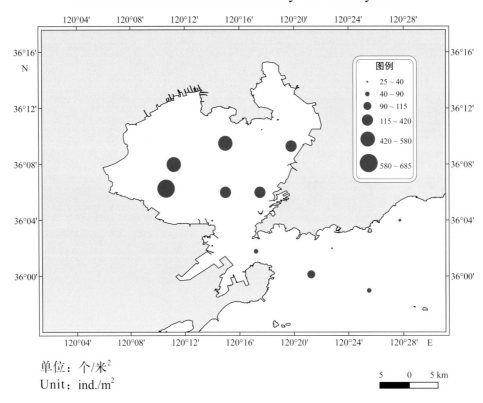

单位：个/米²
Unit：ind./m²

2014年5月底栖生物总密度分布
Distribution of total benthic density in May 2014

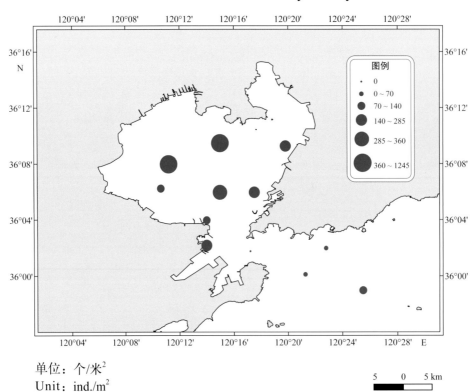

单位：个/米²
Unit：ind./m²

2014年8月底栖生物总密度分布
Distribution of total benthic density in August 2014

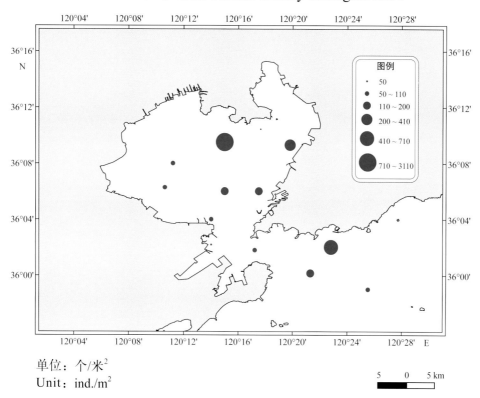

单位：个/米²
Unit: ind./m²

2014年11月底栖生物总密度分布
Distribution of total benthic density in November 2014

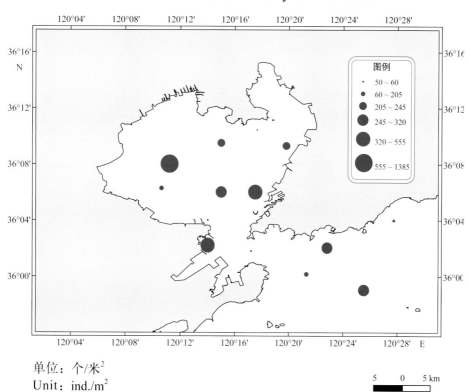

单位：个/米²
Unit: ind./m²

2015年2月底栖生物总密度分布
Distribution of total benthic density in February 2015

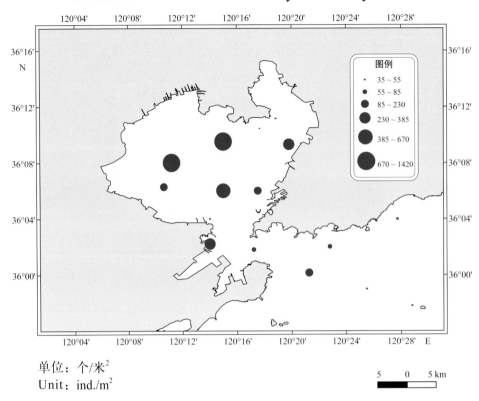

单位：个/米²
Unit：ind./m²

2015年5月底栖生物总密度分布
Distribution of total benthic density in May 2015

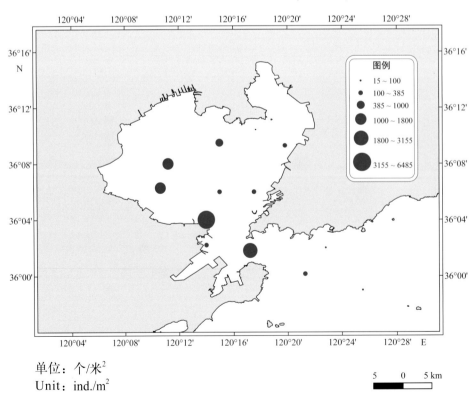

单位：个/米²
Unit：ind./m²

2015年8月底栖生物总密度分布
Distribution of total benthic density in August 2015

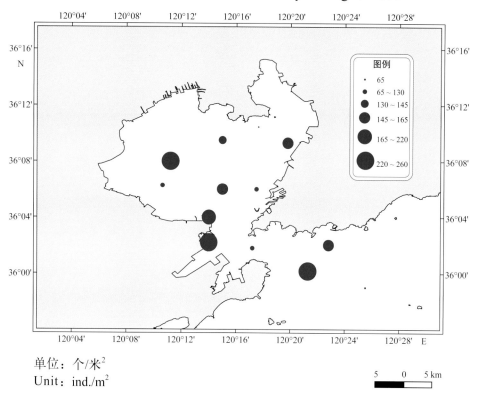

单位：个/米²
Unit：ind./m²

2015年11月底栖生物总密度分布
Distribution of total benthic density in November 2015

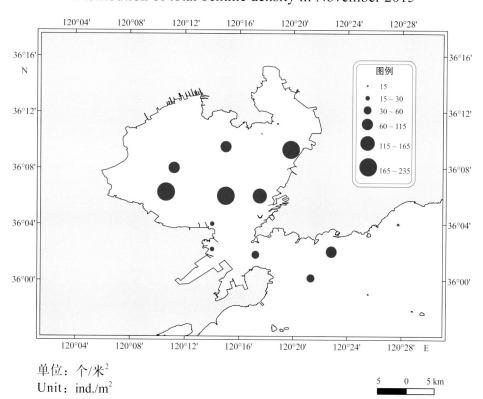

单位：个/米²
Unit：ind./m²